SpringerBriefs in Molecular Science

Chemistry of Foods

Series editor

Salvatore Parisi, Palermo, Italy

AF793674

For further volumes:
http://www.springer.com/series/11853

Ettore Baglio

Chemistry and Technology of Yoghurt Fermentation

Springer

Ettore Baglio
Food Technologist
Catania
Italy

ISSN 2191-5407 ISSN 2191-5415 (electronic)
ISBN 978-3-319-07376-7 ISBN 978-3-319-07377-4 (eBook)
DOI 10.1007/978-3-319-07377-4
Springer Cham Heidelberg New York Dordrecht London

Library of Congress Control Number: 2014940155

© The Author(s) 2014
This work is subject to copyright. All rights are reserved by the Publisher, whether the whole or part of the material is concerned, specifically the rights of translation, reprinting, reuse of illustrations, recitation, broadcasting, reproduction on microfilms or in any other physical way, and transmission or information storage and retrieval, electronic adaptation, computer software, or by similar or dissimilar methodology now known or hereafter developed. Exempted from this legal reservation are brief excerpts in connection with reviews or scholarly analysis or material supplied specifically for the purpose of being entered and executed on a computer system, for exclusive use by the purchaser of the work. Duplication of this publication or parts thereof is permitted only under the provisions of the Copyright Law of the Publisher's location, in its current version, and permission for use must always be obtained from Springer. Permissions for use may be obtained through RightsLink at the Copyright Clearance Center. Violations are liable to prosecution under the respective Copyright Law.
The use of general descriptive names, registered names, trademarks, service marks, etc. in this publication does not imply, even in the absence of a specific statement, that such names are exempt from the relevant protective laws and regulations and therefore free for general use.
While the advice and information in this book are believed to be true and accurate at the date of publication, neither the authors nor the editors nor the publisher can accept any legal responsibility for any errors or omissions that may be made. The publisher makes no warranty, express or implied, with respect to the material contained herein.

Printed on acid-free paper

Springer is part of Springer Science+Business Media (www.springer.com)

Contents

Chapter 1
The Modern Yoghurt: Introduction to Fermentative Processes

Abstract The term 'fermentation' refers to the catalytic transformation of organic substances by microbial enzymes. With reference to fermentation, homofermentative and heterofermentative processes are extensively used in the industry. Fermented milk is a product obtained by milk coagulation without subtraction of serum. The action of fermentative lactic acid bacteria (LAB) is required. Moreover, fermenting agents must remain vital until the time of consumption. The synergic action of selected LAB may be extremely useful: industrial yoghurts show peculiar chemical profiles with relation to lactic acid, main aroma components (diacetyl, acetaldehyde, etc.) and structural polymers such as polysaccharides. Different productive processes are available at present, depending also on the peculiar type of desired yoghurt.

Keywords Acetaldehyde · Acetoin · Acetone · Caseins · Diacetyl · Fermented food · Galactose · Glucose · Lactic acid bacteria · Polysaccharides

List of Abbreviations

ABE	Acetone–butanol–ethanol
D (−)	Dextrogyre
LAB	Lactic acid bacterium
LDB	*Lactobacillus delbruekii* subsp. *bulgaricus*
L (+)	Levogyre
MW	Molecular weight
ST	*Streptococcus thermophilus*

E. Baglio, *Chemistry and Technology of Yoghurt Fermentation*, SpringerBriefs in Chemistry of Foods, DOI: 10.1007/978-3-319-07377-4_1, © The Author(s) 2014

1.1 Fermented Milks: The Peculiarity of Yoghurts

From the historical viewpoint, the preservation of several food products may be obtained with a remarkable improvement of the 'perceived' quality when fermentative processes are used (Leroy and De Vuyst 2004). Actually, the chemical composition of original raw materials, also named 'food ingredients', is crucial. At the same time, fermentative processes should be used for improving microbiological profiles of preserved foods on the basis of marketing requests, consumers' needs and regulatory issues. Anyway, the problem of food safety is the main requirement (Motarjemi 2002). As a consequence, fermentation is one of main techniques for the preservation of food commodities, but the management of many variables is required when speaking of fermented products.

Generally, fermented foods and feeds are subdivided as follows, depending on the origin of main ingredients. A not exhaustive list may be shown here (Campbell-Platt 1995; Pyler 1973; Romano and Capece 2013; Tamime and Robinson 1999; Woolford 1984):

- Fermented milks—yoghurt, *kefir*, etc.—and cheeses
- Alcoholic beverages
- Fermented meats
- Baked foods such as bread, panettone, pandoro, pizza and cakes
- Fermented silages such as silage grass and fish silages.

It should be considered that a large part of 'indigenous' or 'wild' micro-organisms—often simple contaminant microflora—may be used for fermentative purposes with acceptable results. In addition, some positive effect might be obtained in this way against pathogen bacteria in peculiar foods (Zhang et al. 2011). However, there is no assurance that required organoleptic features are always obtained and with acceptable yields. On the other side, many health and hygiene concerns may be discussed.

For these reasons, the industry of fermentative processes has promoted the creation and the use of reliable starter culture: the capability of providing safe and predictable results with a broadened variety of food ingredients is the key of the success in this multifaceted sector (Leroy et al. 2006).

Basically, fermentative processes are managed by means of the correct use of following micro-organisms:

- Lactic acid bacteria (LAB) only. Fermented products: cheeses, yoghurts, sausages, *salami* and silages.
- Yeasts only. Fermented products: alcoholic beverages
- Mixed cultures with LAB and yeasts in synergic association. Fermented products: some wine and baked foods, with the important exclusion of fermented milks.

One of the known and historical fermentative processes concerns the effective preservation of milks. Traditionally, the origin of milk fermentation in Europe is correlated to the appearance of nomadic peoples (Prajapati and Nair 2003). Other examples are known in the ancient China (Liu et al. 2011) or the Eastern Africa (Dirar 1993). With exclusive relation to the diffusion of fermented milks in the European culture, nomadic people were used to preserve milks in containers made from the stomach of animals: the result of this 'fermentative storage' was a dense and acidic food.

In fact, the modern term 'yoghurt' or 'yoghurt' is a corruption of the original Turkish name: *yoghurt* (Prajapati and Nair 2003).

At present, the consumption of fermented milks is very common in many populations with the whole Europe and in other regions. Two different yoghurt typologies may be roughly distinguished:

- Acid milks such as yoghurt and *Kajmac* (Jokovic et al. 2008).
- Acid–alcoholic milks. For example: *kefir*, russian and mongolian *koumiss* types. (Liu et al. 2011; Montanari et al. 1996).

The modern science of fermentation is recent: the conventional date should be coincident with the identification of two main bacterial types—*Lactobacillus delbruekii* subs. *bulgaricus* and *Streptococcus thermophilus*—by the Ukrainian biologist E. Metchnikov at the end of nineteenth century. Because of the effective diversity between Caucasian and European shepherds with relation to the average lifespan, this scientist correlated the higher longevity of eastern animals with the peculiar diet and the abundant consumption of fermented milks (Pot and Tsakalidou 2009).

In 1906, the company '*Le Ferment*' began to sell a fermented milk in France. The original brand name—*Lactobacilline*—was correlated with the use of selected LAB according to Metchnikoff's suggestions and techniques. As a result, the initial success on the market of milk products allowed the word 'yoghurt' to enter in the common language: the *Petit Larousse* presented this term as a common word.

1.2 Fermentation and Processes

Generically, the term 'fermentation' refers to the catalytic transformation of organic substances, mainly carbohydrates, by enzymes of microbial origin (Cappelli and Vannucchi 1990). These modifications may represent some undesired alteration; on the other hand, the action of microbial enzymes by selected micro-organisms may be used for the safe and convenient production of food products. Table 1.1 shows a small selection of different life forms with industrial applications.

Industrial fermentative processes for food applications may be approximately subdivided in two categories:

(1) Homofermentative processes. The production of a single compound is obtained. For example: alcoholic fermentation; obtained product: ethyl alcohol
(2) Heterofermentative processes. Two or more final products are obtained. Example: the acetone–butanol–ethanol (ABE) fermentation can be used with the aim of producing acetone, ethyl, isopropyl and butyl alcohols (Park et al. 1989).

Table 1.1 A selection of fermentative micro-organisms for the production of yoghurts and other fermented foods (De Noni et al. 1998; Simpson et al. 2012)

Organism	Type	Main food applications
Saccharomyces cerevisiae	Yeast	Wines, beers, baker's yeast, wheat and rye breads, cheeses, vegetables, probiotics
Saccharomyces bayanus	Yeast	Fermented milks
Streptococcus thermophilus	Yeast	Yoghurt, hard and soft cheeses
Lactobacillus bulgaricus sub. *delbrueckii*	Bacterium	Yoghurt
Propionibacterium shermanii	Bacterium	Swiss cheese
Lactobacillus casei	Bacterium	Cheeses, meats, vegetables, probiotics
Gluconobacter suboxidans	Bacterium	Vinegars
Penicillium roquefortii	Mould	Gorgonzola cheese
Penicillium camembertii	Mould	Camembert and brie cheese
Aspergillus oryzae	Mould	Soy sauce, sake
Candida famata	Yeast	Meats

Fermentative micro-organisms can be bacteria or fungi. For example, several useful bacteria belong to *Lactobacillus, Clostridium, Nitrobacter* and *Acetobacter* genera. With reference to fungi, most known life forms with industrial importance are yeasts and moulds.

Environmental conditions affect the survival of micro-organisms and the duration of related fermentations; consequently, several fermentative processes may be stopped because of the inhibitive action of main fermentation products. For instance, the alcoholic fermentation is stopped if the percentage of produced ethyl alcohol reaches 14–16 %. Anyway, main fermentative processes are related to the transformation of carbohydrates. Five typologies may be described here as follows:

- Alcoholic fermentation
- Homolactic fermentation
- Heterolactic fermentation
- Propionic fermentation
- Butyric fermentation
- Oxidative fermentation
- Citric fermentation

1.2.1 Alcoholic Fermentation

This complex process is mainly carried out by yeasts such as *Saccharomyces* genus. Chemically, two different substrates may be fermented:

- D-glucose, also named dextrose, corn or grape sugar. Chemical formula: $C_6H_{12}O_6$, molecular weight (MW): 180.16 g mol^{-1}
- D-fructose, also named 'fruit sugar', levulose. Chemical formula: $C_6H_{12}O_6$, MW: 180.16 g mol^{-1}.

These simple molecules can be found in grapes, barleys and wheats. However, the preventive hydrolysis of ring structures is required before the real fermentation process. For example, glucose is hydrolysed (glycolytic reaction) and the resulting pyruvate is decarboxylated to acetaldehyde. Subsequently, acetaldehyde is reduced to ethyl alcohol. Final products are ethyl alcohol and carbon dioxide. However, other by-products may be obtained: glycerine, various organic acids, etc.

1.2.2 Homolactic Fermentation

This process, mainly carried out by acid-forming *Lactobacillus* and *Streptococcus* bacteria, corresponds to the microbiological transformation of glucose to lactic acid by the reduction of pyruvic acid. Because of the high efficiency of the fermentative process, the homolactic strategy is useful for the preparation of yoghurts, the ripening of cheeses and the preservation of several vegetables (Fleming et al. 1985).

1.2.3 Heterolactic Fermentation

Differently from homolactic fermentation, this process is not specific with reference to final products: lactic acid, ethyl alcohol and carbon dioxide are obtained at the same time (fermented substrate: glucose). This time, involved bacteria belong to *Leuconostoc* genus. Most known and studied applications concern the production of acid–alcoholic milks such as *kefir* (Lyck et al. 2006).

1.2.4 Propionic Fermentation

This process is carried out by *Propionibacterium* micro-organisms. The initial substrate is lactic acid, and final products are propionic acid, acetic acid and carbon dioxide. Because of the notable production of volatile acids, the propionic fermentation is mainly used for the 'maturation' of *Emmentaler* cheeses (Benjelloun et al. 2005).

1.2.5 Butyric Fermentation

Actually, this process is not desired in the industry of fermented and normal products because of the occurrence of unpleasant and unwanted substances. For example, the so-called delayed swelling of some cheese is considered such an important alteration (McSweeney 2007). Involved micro-organisms are butyric bacteria including *Clostridium tyrobutiricum* in particular. The end product of this fermentation is butyric acid, but other compounds—acetic acid, carbon dioxide and hydrogen—are also obtained. The fermentative pathway concerns the initial glycolysis of glucose with the production of two pyruvate molecules (Sect. 1.2.1). Subsequently, pyruvates are turned into acetyl coenzyme A by means of an enzymatic oxidative process. Finally, acetyl coenzyme A is converted into butyryl phosphate after a four-step enzymatic reaction (Duncan et al. 2002).

1.2.6 Oxidative Fermentation

This multifaceted process can be carried out by either obligate or facultative aerobic micro-organisms in presence of oxygen. The oxidation of the peculiar substrate gives the final production of carbon dioxide and water, although different products may be obtained if the demolition is partial. An example is the conversion of ethyl alcohol to acetic acid by the action of *Gluconobacter* and *Acetobacter* micro-organisms. This transformation, normally associated with the production of vinegar, can occur spontaneously in wines when the absence of oxygen is not assured.

1.2.7 Citric Fermentation

This peculiar type of fermentation is carried out by *Aspergillus niger.* The process is usually observed in soils with remarkable amounts of carbohydrates; however, the quantity of trace elements such as iron, copper and anions like phosphates should be negligible. In these conditions, the Krebs cycle is altered with the consequent accumulation and excretion of citric acid (Max et al. 2010). The notable yield of produced citric acid has determined the wide use of this pathway in the industry (Roukas 1991).

1.3 Fermented Milks and Yoghurts

At present, fermented milk products may correspond to a wide variety of different typologies, depending on the result of environmental conditions, used micro-organisms and productive processes. The common point is the demonstration of an intense lactic fermentation due to the development of LAB into milks. Maybe, the association of LAB with other co-fermenting life forms—yeasts, acetic acid bacteria and moulds—is observable with different results. However, fermented milks should maintain a constant and acceptable quality from the viewpoint of normal consumers when fermentative processes are pre-designed with the aim of producing a small number of end products, mainly lactic acid.

By a general viewpoint, fermented milk is a product obtained by milk coagulation without subtraction of serum (Corradini 1995). The action of fermentative microorganisms is required and should exclude other coagulating or gelling processes. Anyway, fermenting LAB must remain vital until the time of consumption. The classification of fermented milks can be made on the basis of fermenting microbes (Corradini 1995):

- Thermophilic acidic milk. The main product of fermentative reactions is lactic acid. Required thermal range for fermentation is 37–45 °C
- Mesophilic acidic milk. The main product of fermentative reactions is lactic acid. Required thermal range for fermentation is 20–30 °C
- Acidic alcohol milk. Main end products of fermentative reactions are lactic acid, ethyl alcohol and carbon dioxide. Required thermal range for fermentation is 15–25 °C.

Among these categories, the group of thermophilic acidic milks has been constantly evolving in last decades with a remarkable augment of market revenues. The first and probably best known sub-type of thermophilic acidic milks appears to be the 'European' yoghurt. Interestingly, the diffusion of yoghurts is now observed worldwide in spite of the 'regional' tradition of the ancient *yoghurt* food (Sect. 3.1).

Yoghurt, also named 'yogurt', is a product made from heat-treated milk. It has to be considered that the original 'raw material' may be homogenized before the addition of LAB cultures containing *Lactobacillus bulgaricus* and *Streptococcus thermophilus* (Chandan and Kilara 2013). Similarly, yoghurt can be defined as the product of the lactic fermentation of milks by addition of a starter culture, with the consequent decrease of pH to 4.6 or lower values (Tamime 2002). By the viewpoint of industrial processors, yoghurts can be subdivided into two types.

First of all, a 'set-style' yoghurt is made in retail containers with the necessity of giving a continuous and undisturbed gel structure in the final product (Tamime and Robinson 1999). On the other hand, the 'stirred' yoghurt has a delicate protein-made gel structure: this network is reported to develop during fermentation (Benezech and Maingonnat 1994). With exclusive relation to stirred yoghurt manufacturing processes, gel networks are disrupted by stirring before mixing with fruit; subsequently, the stirred fluid is packaged. Stirred yoghurts should have a smooth and viscous texture (Tamime and Robinson 1999). In terms of rheology, this food corresponds to a viscoelastic and pseudoplastic product (De Lorenzi et al. 1995).

The matter of texture can be used as a discriminator feature: yoghurts are available in a number of textural—liquid, set, and smooth—types. In addition, commercially available yoghurts may be present with different amounts of declared fat contents. Finally, the possibility of flavour additives can suggest the production of 'natural', fruit- and cereal-enriched products.

On the other side, yoghurts may be consumed alone, as a snack or part of a complex meal, as a sweet or savoury food. They can be used from starters to desserts, from meat or fish dishes with the aim of innovating culinary traditions.

Differently from other food commodities of the recent—and not industrialized—past, yoghurts are virtually available during the whole year without scarcity periods. This versatility and the recognized acceptability as a healthy and nutritious food have progressively determined the widespread popularity of this peculiar fermented milk across all population subgroups (McKinley 2005). The resting part of fermented milks is apparently multifaceted: these products may be defined as non-traditional foods, made by means of homofermentative processes and microbial strains, with relatively high optimum temperatures. With concern to these foods, it should be also clarified that used microbial cultures are used singularly or in combination between them and in association with mesophilic strains also (thermal range: 20–30 °C).

Basically, the general 'lactic' term means all food preparations with the presence of selected bacteria. These life forms should be recognized able to rebalance the intestinal microflora (Savini et al. 2010) and produce specific inhibitory substances (bacteriocins) and other metabolites that are also active towards pathogenic micro-organisms (Flynn et al. 2002).

By the consumeristic viewpoint, the commercial success of fermented milks seems to be correlated with the well-known 'sour taste'. In other words, fermented foods appear refreshing in taste and greatly appreciated in hot climates. Similarly to bottled colas and fruit juices, several of these products may also show peculiar features: a famous example is the Caucasian *kefir* because of the known effervescence and other co-factors: notable homogeneity, creaminess, etc. (Simova et al. 2002). For these reasons, the penetration of fermented milks appears to be virtually extended without boundaries because of the number of different types, combinations and presentations.

The global yoghurt market is expected to surpass $67 billion by the year 2015 (Global Industry Analysts 2010). The above-mentioned prediction is also favoured by the increasing popularity of yoghurts as functional foods. The rapid growth of the global dairy industry is attributed mainly to the advent of functional products: features such as low-sugar, low-fat, cholesterol-lowering and favourable impacts on the digestive health appear to be convincing and attractive arguments. Precise marketing strategies need reliable technologies and the profound knowledge of chemical transformations in the initial raw material, intermediates and the final product. Consequently, a chemical perspective is needed.

1.4 Features of Lactic Microflora in Yoghurts and Related Chemical Profiles

The LAB heterogeneous group is able to ferment various substrates with the consequent production of numerous products of interest for the food industry. Basically, these micro-organisms have following general features.

LAB are definable as Gram-positive, catalase-negative bacteria with different shapes and associations (Stiles and Holzapfel 1997). They may be found arranged in chains of two or more elements; generally, there is no risk of suspect pathogenicity. These life forms can grow up on culture media as small and colourless colonies. From the nutritional viewpoint, LAB are well known for their limited biosynthetic capacity: as a result, a considerable bioavailability of vitamins, amino acids and nitrogen bases is needed. Moreover, LAB are generally unable to reduce nitrate ion to nitrite with some peculiar exception.

With relation to environmental conditions, LAB are recognized as oxygen-tolerant anaerobic bacteria: the necessary energy is obtained through the phosphorylation of the substrate. For this reason, LAB show a fermentative energy metabolism and are able to produce lactic acid from one or more carbohydrates through the homolactic or heterolactic way. According to the Bergey's Manual of Systematic Bacteriology, the classification in subgroups is justified on the basis of the preferred or demonstrated fermentative pathway (Kandler and Weiss 1986; Schleifer 1986):

(a) Obligated homofermentative bacteria. Glucose is entirely transformed into lactic acid via the Embden–Meyerhof glycolytic pathway
(b) Facultative heterofermentative bacteria. These life forms are homofermentative bacteria with the ability of encoding an inducible phosphoketolase (Lindgren and Dobrogosz 1990). On these bases, they are able to ferment pentoses with the production of lactic and acetic acids. On the other side, hexoses are fermented in the homolactic way
(c) Obligated heterofermentative bacteria. These life forms do not have a key enzyme in the glycolytic pathway (targeted molecule: fructose 1,6-diphosphate). For this reason, they are accustomed to ferment glucose according to the 6-phosphogluconate way with production of lactic acid in equimolar ratio, carbon dioxide and ethyl alcohol or acetic acid.

Optimal thermal ranges for growth vary from 15–20 °C to 40–45 °C. Actually, some species is known to grow at 4 °C while other micro-organisms may arrive up to 50–55 °C. In addition, the resistance to thermal treatments (pasteurization) has been observed in several ambits and foods (Franz and Von Holy 1996). The presence of LAB in raw milks and derivatives is explainable because of the adaptability in a variety of environments, from vegetables to the digestive tract of some mammals. As a result, soils and superficial waters may be found with living LAB because of the prior contamination from animals or plants.

The industrial importance of LAB in the industry of fermented foods is dependent on the demonstrated ability to produce various useful substances (Buckenhüskes 1993). In general, the use of LAB is widely observed and reported with concern to the industrial and artisanal production of cheeses (including industrial curds for subsequent reworking), fermented milks, meats, fermented silages and vegetables and bakery (Foschino et al. 1995; Ottogalli 2001; Volonterio Galli 2005).

As already mentioned, yoghurt is the combined result of the development of *Streptococcus thermophilus* (ST) and *Lactobacillus delbruekii* subsp. *bulgaricus* (LDB). These LAB are thermophilic homofermentative micro-organisms: the result of the whole fermentative process is exclusively lactic acid.

Differently from other opportunistic associations, the synergic interaction between ST and LDB is extremely efficient: when speaking of yoghurts, the acidification of the food medium (raw milk) is concomitant with the formation of new aromas. This element is extremely important because of (a) the commercial and technical classification of the fermented product and (b) the production of polysaccharides.

The role of streptococci and lactobacilli in the yoghurt manufacture can be summarized as follows: (a) milk acidification, (b) synthesis of aromatic compounds and (c) development of desired texture and viscosity. The evaluation of the final aroma is generally based on the production of acetaldehyde, a major aromatic compound of yoghurt, whereas the thickening character is based on measurements of milk viscosity (Bouillanne et al. 1980; Zourari et al. 1992).

The above-mentioned synergicity of streptococci and lactobacilli is well demonstrated with reference to the production of aldehydic compounds: associated ST and LDB can produce more acetaldehyde than the sum of produced amounts by separate fermentations. It has been reported that the synergistic association can produce 22–25 ppm of acetic aldehyde after 4 h of incubation, while LDB is able to obtain only 10 or 11 ppm in the same condition and ST does not exceed 3.0 ppm (Battistotti and Bottazzi 1998). Anyway, the content in acetaldehyde appears to range from 20 to 50 ppm: in addition, it seems to remain typically constant during the storage of fermented products. Acetic aldehyde is generally associated with 1–4 ppm of produced acetone, 2.5–3.5 ppm of acetoin and 0.5–1.0 ppm of diacetyl when speaking of commercial yoghurts.

With reference to the above-mentioned synergicity, the activity of specific enzymes for acetaldehyde and other catalytic reactions appears similar when the two different micro-organisms are compared (Battistotti and Bottazzi 1998). Interestingly, the observed absence of the enzyme α-carboxylase in both micro-organisms has suggested that acetaldehyde cannot be derived by pyruvic acid—this is the normal fermentative way—while the enzymatic activity of threonine aldolase is reported for LDB. Finally, the stability of acetic aldehyde during the storage period of yoghurt seems to be dependent from the absence of the enzyme alcohol dehydrogenase in both species.

On the other side, the use of *Lactobacillus acidophilus*, probiotics such as 'acidophilus milk' and related preparations appears unsatisfactory with relation to obtained aromas: in fact, the quantity of the produced acetaldehyde (and the consequent flavour) is very low or negligible when used species have alcohol dehydrogenase, differently from LDB.

The main role of ST and LDB in the yoghurt manufacture concerns the acidification of milks by means of the production of lactic acid from lactose. It is known that main milk proteins—caseins—tend to make notable agglomerations starting from a pH of 5.2–5.3 at room temperature. The best coagulating effect occurs when the isoelectric point of casein is reached (pH 4.6): in these conditions, dispersed casein micelles can make good agglomerations depending on environmental conditions—minimum temperature: 10 °C—because of the deficiency of calcium phosphate. Actually, the ratio between soluble salts and the insoluble calcium phosphate seems to have a decisive influence on the coagulation of gel networks.

The pH of milk can reach 4.8–5.0 log units at temperatures of about 30 °C by addition of acidic solutions and/or by means of the simple lactic acid fermentation. As a consequence, low pH values cause the decrease of the ionization of acid functions on caseins and the reduction of measurable redox potentials. In other terms, the solubility of calcium salts in the aqueous matrix of the milk is notably favoured and increased when pH is lowered. Because of the original placement of calcium ions on the surface of phosphocaseinate micelles, casein chains gradually suffer a remarkable demineralization: 50 % of colloidal calcium is dissolved when pH ranges from 5.7 to 5.8. It should be noted that visible textural modifications of rheological features are generally observed at this point (Trejo 2012) while the dissolution of calcium ions becomes complete at pH <5.0. Probably, the association between proteins appears to be mainly caused by saline bonds when pH is 4.6 (Lucey and Fox 1993).

A profound disorganization of casein micelles is produced during the acidification process with the concomitant modification of spatial arrangements. An important neutralization of electric charges is verified at the isoelectric pH with the progressive decrease in hydration: this complex series of physicochemical variations determines usually the insolubilization of caseins. The final clot can be seen as an aggregate of solubilized proteins absorbed in their aqueous matrix; the great fragility of the lactic acid clot is caused by the electrostatic and hydrophobic nature of existing bonds in the micellar state. Three main factors seem to influence the nature of the acidic clot: the clotting temperature, the rate of acidification and the concentration of proteins. The higher the amount of caseins, the higher the consistency of the resulting mass.

The rate of acidification is crucial for the structure of the clot: rapid or very rapid rate values lead to an unstructured and flocculent clot, while slow acidification appears to determine a properly structured mass.

Fundamental differences between acidic clots obtained by acidification can be explained in this way: anyway, viscous masses may be obtained by adding mineral substances or organic acids in high concentration to the original milk. Should the first approach be used, the resulting clot would be usually fragile but uniform; in the second situation, the isoelectric point will determine the remarkable collapse of proteins with the consequent release of water. Actually, the mechanism is not completely known at present by the chemical viewpoint: it may be inferred that the main critical factor is the time of acidification. This quantity is definable as the required time in the fermentation process for the production of minimal amounts of lactic acid and the consequent displacement of calcium ions without the clear alteration of electrical and hydrophobic micellar charges.

On the other side, the phenomenon of the acidic transformation of proteins can prevails on the observable shift of mineral salts when concentrated acids are massively added: as a consequence, casein micelles tend to flocculate irreversibly (Tuiner and De Kruif 2002).

Anyway, lactic acid reduces the pH of the milk and causes the progressive solubilization of the micellar calcium phosphate. In other terms, the demineralization and the destabilization of casein micelles are produced with the consequent and

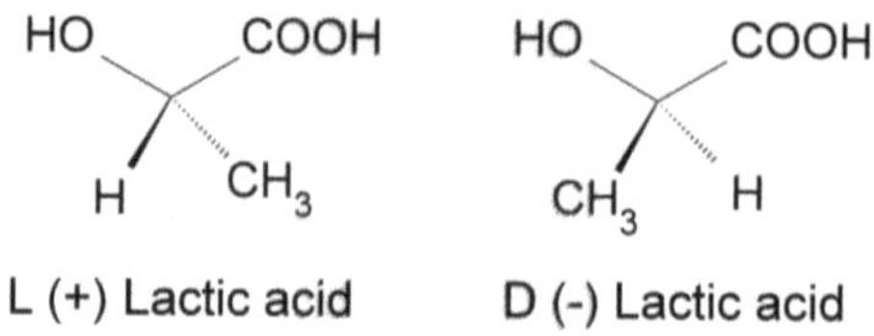

Fig. 1.1 Optically active isomeric forms of lactic acid: L (+) and D (−) structures. BKchem version 0.13.0, 2009 (http://bkchem.zirael.org/index.html) has been used for drawing this structure

complete precipitation of caseins in a pH range between 4.6 and 4.7 (Fox 2008). In addition, lactic acid is critical with relation to sharp, acid tastes and the resulting aroma of produced yoghurts.

On the other side, the excessive acidification may affect organoleptic properties of the final product. This undesired failure may depend on used LAB strains, lactobacilli above all (Accolas et al. 1977; Bouillanne et al. 1980). With relation to the homolactic fermentation of lactose, two optically active isomeric forms of lactic acid are obtained: the levogyre L (+) structure is produced by ST and the dextrogyre D (−) form is obtained by LDB. It should be also noted that the development of the two species in yoghurts is also influenced by the availability of formic acid. The ratio between the two forms in the racemic mixture depends on the intensity of development of the two bacterial species, although the accumulation of the levogyre isoform by some lactobacilli may induce a specific racemase. Should this situation be verified, the conversion of the levogyre structure in the dextrogyre isoform would be observed until the final equilibrium is obtained (Narayanan et al. 2004) (Fig. 1.1).

Generally, the amount of L (+) lactic acid is between 50 and 60 % of the racemic mixture. The total concentration of this acid in yoghurts appears to be in the 0.7–1.2 % range, while pH values are between 3.9 and 4.2 (De Noni et al. 1998).

With relation to first steps of the homofermentative process, the initial substrate—lactose—is transported into cells by means of a dedicated permease. The subsequent and absolutely needed step is the hydrolytic separation of the disaccharide in glucose and galactose by the specific β-galactosidase. The glucose is rapidly phosphorylated, turned into two triosephosphates by means of the aldolase enzymatic system and finally converted to pyruvic acid according to the simple glycolytic way (Sect. 1.2.1). Pyruvic acid is finally turned into lactic acid by the specific lactate dehydrogenase enzyme. On the other side, galactose is discharged outside the cell without fermentation (Battistotti and Bottazzi 1998).

With reference to industrial and artisanal yoghurts, another aspect of technological interest concerns the production of polysaccharides. In fact, LAB species such as ST and LDB can produce polysaccharides when developed in milk: chemically, the structure of these carbohydrates is based on galactose and glucose. Produced amounts are reported to be higher when LAB synergic species are in association: 800 mg/l of milk, while ST can produce up to 350 mg/l and LDB may arrive to 425 mg/l depending on the peculiar strain. The importance of polysaccharides is purely structural because polymers are obtained in form of filaments: these quasi-linear structures can bind cells together. Consequently, clots of coagulated casein may finally show appreciable resistance against syneresis—the expulsion of fluid masses or whey from a structured but chaotic network—and the consequent uniformity (Everett and McLeod 2005).

The production of polysaccharides is influenced by many factors, including temperature; generally, 0.2 % of the total weight is composed of polysaccharides after 10–15 days in commercially available yoghurts. This concentration has a positive effect on the structure of the product that appears smooth and fine on the palate (Battistotti and Bottazzi 1998).

1.5 Industrial Yoghurts: Preparation of Milks

By the commercial viewpoint, yoghurt types may be identified as follows (Sect. 3.1):

- White yoghurts. Ingredients: milk with the possible addiction of milk creams
- Dessert-type yoghurts. These products contain also fruit pieces, puree or juice, herbs or other ingredients: sugar, cereals, cocoa, malt, chocolate, royal jelly, honey, coffee and other vegetable juices
- Enriched yoghurts. In other words, these foods are 'plain' (white) or dessert-type yoghurts with mineral substances, vitamins, oligosaccharides, fibres and/or other functional ingredients or probiotics.

Generally, these yoghurts are produced in skim or whole types depending on the fat content in the finished product up to 1 % or more than 3 %, respectively. However, partially skimmed products can be prepared.

Basic ingredients, used milks above all, have to be carefully evaluated before the production. First of all, the complete absence of residues of antibiotics and detergents has to be confirmed because of the known sensitivity of LAB even at low levels of contamination. In detail, the presence of synthetic detergents and antibiotics can determine the dangerous slowdown of the lactic fermentation with consequent low acidification. Another possible danger is the excessive extension of processing times.

Therefore, milks containing antibiotic residues or detergents are generally avoided for the production of yoghurts. Moreover, the amount in proteins in the original milk is extremely important: high values contribute significantly to the formation of creamy and syneresis-resistant yoghurts (Tamime and Robinson 1999). It may be inferred that processing costs depend strongly from the amount of proteins in the intermediate clot: this number should be ranged between 3.8 and 3.9 %. For this reason, the basis should be an initial concentration of nitrogen-based molecules between 3.0 and 3.4 %.

Even the microbiological quality must be excellent, especially with concern to the estimation of heat-resistant micro-organisms and spores. In fact, high microbial contaminations are often associated with the presence of enzymes capable of producing sensorial and textural alterations. The development of psychotropic microflora such as *Pseudomonadaceae* can be considered the cause of the detection of thermostable proteases and lipases: these enzymes may be not inactivated after normal heat treatments (De Noni et al. 1998). Should this situation be verified, the following proteolytic degradation could determine the alteration of creamy textures with consequent serum separations. In addition, rancid tastes may be observed after the hydrolysis of triglycerides. Proteolytic enzymes may also result from the lysis of somatic cells. For these reasons, the recommended level should be <300,000 cells/ml (De Noni et al. 1998; Ruegg 2005).

Finally, the absence of bacteriophages is highly recommended: these life forms may be highly resistant against sanitization procedures. Clearly, should their presence be demonstrated in the initial milk, the fermentative pathway could be completely modified in comparison with predictable reactions and obtained results, in terms of pH, acidification and production of polysaccharides.

After the selection and usual quality controls on raw milks, the subsequent step is the pasteurization of a mixture consisting of milk and added fats, proteins, sugar or other ingredients (where possible). This mixture is then inoculated with specific LAB culture: the fermentation process can finally be carried out. Obtained yoghurts may receive the addition of flavourings, fruit preparations and other food additives for specific functions (Sect. 3.1).

It has to be considered that raw milks cannot be used immediately: first of all, a sort of physical removal of foreign substances has to be conducted by mild centrifugation. This process is apparently preliminary and without chemical consequences: however, the centrifugation should be carefully performed because of possible risks in subsequent steps.

In fact, 'purified' milks cannot be pasteurized without the preliminary correction of fat contents through a complete skimming and the addition of fatty creams. At the same time, this correction determines the quantitative variation of proteins and the final consistency of intermediate milks: the aim is substantially correlated with the necessity of producing stable creams without phase separations and syneresis (expulsion of whey).

As a result, lipids can vary between 0.1 % and 3.0–3.5 % in low-fat and whole yoghurts, respectively (Tamime and Robinson 1999). Other additions are possible before pasteurization: for instance, the preparation of sweetened—fruit or flavoured—yoghurts may require the use of artificial sweeteners, glucose and fructose (grape sugar), fructose only, etc. (Sect. 3.1).

The amount of added substances and food additives may depend on the sweetness, also defined sweet power, of used sugars: anyway, it should not exceed 10 %. In fact, the development of LAB cultures in milks and consequent acidification rates may be notably slowed down due to excessive concentration of dissolved sugars and resulting osmotic pressure values (Dalla Rosa and Giroux 2001; Tamime and Robinson 1999).

With exclusive concern to the problem of tastes, the glucidic content in commercial yoghurts is generally determined by the detectable amount of sugars in semi-finished fruits and the expectation of normal consumers. Ingredients such as malt, cocoa and cereal flours may be also added; the same thing can be affirmed when speaking of pre-biotic substances such as inulin, a polysaccharide of vegetable origin with the interesting property of favouring the colonization of useful microflora in the human intestine. These constituents are added with the aim of promoting the efficient dissolution in subsequent processing steps—heating and homogenization—before pasteurization (Tamime and Robinson 1999). During this step, the milk can also be vigorously agitated without causing damage to clot structures: on the other side, above-mentioned ingredients might be added at the end of fermentation, and the continuous agitation could be dangerous in this step.

After the addition of sugars at least, the protein content of raw milks is also corrected for promoting good resistance to syneresis and acceptable rheological properties for the resulting product (Tamime and Robinson 1999). Generally, the content of proteins in the final yoghurt should be ranged between 3.8 and 3.9 % by means of milk concentration or the simple addition of proteins to the original raw materials. The first system—water evaporation—is carried out by heating the mixture up to 75–90 °C under vacuum: 15–20 % of the total quantity of water can be removed in this way. On the other hand, it should be remembered that rheological properties of yoghurts are also correlated with fat and protein contents; as a result, the adequate consistency should be obtained when fat content is <0.5 %, and the amount of proteins is similar to 5.0 %. For these reasons, 35–40 % of the initial water in raw milks should be eliminated (Tamime and Robinson 1999).

However, the process of evaporation may be expensive if above-mentioned objectives are compulsory: consequently, the concentration is often obtained by ultrafiltration. Alternatively, the second approach—the simple addition of dried milk proteins—is used.

Substantially, ultrafiltration is performed by means of the use of membrane filters with pores of fixed dimensions: the principle of the procedure aims to filter and eliminate inorganic ions, organic acids and lactose by simple size exclusion, similarly to the chromatographic technique of size exclusion (Fig. 1.2).

Anyway, the result of milk correction is a complex biphasic mixture: an agglomeration of fat matters and proteins—the retentate—is dispersed in the so-called permeate, the aqueous solution of salts and various sugars. However, the expected loss of calcium and phosphorus is related to the soluble fraction: about 60 and 50 % of original calcium and phosphorus contents, respectively, are still bound to casein chains and consequently 'blocked' in the retentate (McMahon and Oommen 2013; Uricanu et al. 2004). Milk proteins can be added as mixed dried powders, rennet caseins, whey powders and ultrafiltered proteins. In particular, whey powders and ultrafiltered milk proteins are certainly more expensive than powdered milk or caseinates: on the other side, their addition allows to obtain excellent products with concern to the final creaminess.

By contrast, the addition of protein powders is always associated with the often perceived sensation of 'grittiness': this defect is determined by the incomplete dissolution of powders into the final medium. The same failure can be observed in other dairy products. It can be affirmed that evaporation or ultrafiltration do not show similar defects: in detail, vacuum evaporation is the best procedure because of good results with reference to the milk 'normalization' or correction; moreover,

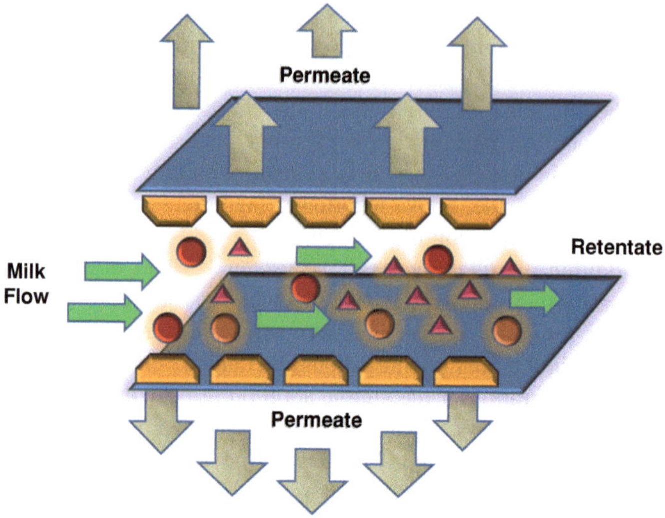

Fig. 1.2 The process of ultrafiltration for the production of yoghurts. Milky mixtures are forced to pass through membranes filters. The procedure aims to filter and eliminate inorganic ions, organic acids and lactose by simple size exclusion, similarly to the chromatographic technique of size exclusion

the concentration of air in the milky mixture is remarkably reduced (Tamime and Robinson 2007) with the consequent increase in lactic acid by LAB fermentation. Other advantages are related to:

- The necessity of avoiding the germination of *Bacillus* spores during the fermentation
- The formation of more homogeneous clots
- The remarkable reduction of oxidative processes on certain vitamins
- The elimination of short-chain fatty acids and other substances may confer abnormal tastes and aromas to the final yoghurt.

Consequently, the minimization of air bubbles in the milky mixture is compulsory and required (Tamime and Robinson 1999–2007).

After correction, the complex mixture has to be homogenized with the aim of reducing the size of fat globules during the fermentation. The homogenization determines also the interaction between triglycerides and proteins on the one side and phospholipids on the other side; last molecules are obtained from the rupture of fat globules. The hydrophilicity is notably increased in spite of the hydrophobic nature of fats; as a consequence, the intermediate clot tends to be resistant against syneresis and the danger of increased creaminess. In addition, the augment of globular surfaces after homogenization determines also the peculiar white colour of clots because of the enhanced light reflection (Petridis et al. 2013).

With reference to this aspect, it should be also remembered that the milky mixture contains still a remarkable amount of calcium ions with added light

Table 1.2 Observed modifications of microbiological and physicochemical profiles in yoghurts and technological causes (De Noni et al. 1998; Simpson et al. 2012)

Reported phenomena	Observed Effects
Microbiological events	
Destruction of pathogenic micro-organisms	• Enhanced sanitization
Destruction of vegetative competitive microflora	• Rapid fermentation and preservation
Chemical reactions	
Inactivation of the most part of natural antibacterial substances	• Rapid fermentation
Inactivation of microbial lipases and proteases	• Increased stability of the taste and texture
Caramelization of lactose with the formation of formic acid	• Stimulation of the growth of lactic bacilli
Activation of the Maillard reaction	• Formation of aromatic compounds
Chemical and physical reactions	
Interaction between caseins and whey-denatured proteins	• Improvement of the consistency and rheological properties of clots • Reduction in the tendency to syneresis
Lowering redox potentials for the removal of oxygen and the liberation of sulphuric groups	• Rapid fermentation and enhanced stability towards oxidation

reflection. The same phenomenon is visible on the surface of certain cheeses (Parisi et al. 2009).

After homogenization, the subsequent step is the pasteurization of milky mixtures at 85–90 °C (time: 10–30 min) by means of heat exchangers or 'shell and tube' plates. Heat treatments have two main roles: the first and well-know reason is naturally the necessity of eliminating contaminant and pathogen agents. However, all possible thermal processes can also produce several modifications of microbiological and physicochemical profiles: these variations may be useful when milky mixtures have to be subjected to fermentation and preservation techniques. A selection of thermal effects on milks and milk derivatives is shown in Table 1.2.

When speaking of milk and milk derivatives, the most important technological effect appears connected to the interaction between casein chains and whey proteins through the formation of hydrophobic bonds and disulphide bridges. These concomitant factors may determine a greater hydration of micellar caseins and the formation of viscous clots in a subsequent stage with low tendency to syneresis. It has to be noted that traditional yoghurts require more drastic heat treatments if compared with industrial products: however, pasteurization processes may cause severe nutritional changes. For instance, the complex of Maillard reactions has to be carefully evaluated because of the reduction in bioavailable lysine (1–5 %), degradative reactions of lipids and carbohydrates and the decrease in some water-soluble vitamins (Pizzoferrato et al. 1998).

At the end of the pasteurisation step, the milk mixture is cooled to 40–45 °C and inoculated with 1:1 or 2:1 mixed cultures of ST and LDB (De Noni et al. 1998).

1.6 The Lactic Inoculum

The duration of fermentation processes depends on properties of used strains, the physical state of the microbial mixture (liquid or lyophilized culture) and the desired level of acidity in the final yoghurt. By a general viewpoint, 3 h at least are required while the recommended maximum duration should be 9 h. By contrast, longer fermentation times may be allowed in the production of yoghurts with low acidity if combined with lower temperatures: related conditions should be 15 ± 3 h at 33 ± 2 °C. Should these parameters be respected, the lactic fermentation would be easily controlled and suddenly stopped when the desired degree of acidity is reached. The most significant phenomenon during the fermentation process concerns the transformation of lactose, $C_{22}H_{12}O_{11}$, into lactic acid, $C_3H_6O_3$, and galactose. The chemical balance of the fermentation process is as follows:

$$C_{22}H_{12}O_{11} + H_2O \rightarrow 2C_6H_{12}O_6 \rightarrow 2C_3H_6O_3 + C_6H_{12}O_6$$

Normally, the final amount of lactic acid in yoghurts is between 0.8 and 1.3 %: this quantity determines substantially low pH values (4.0–4.5) because this fermentative pathway does not contemplate other sub-processes and consequent by-products.

As discussed in Sect. 1.4, two isomeric forms of lactic acid may be found. D (−) lactic acid might have some nutritional significance: this stereoisomer is difficultly metabolized. Anyway, 20–40 % of the total amount of the original lactose is converted into lactic acid with reference to yoghurts while the remaining disaccharide does not exceed 5.5 %. This quantity is not negligible; however, the importance of yoghurts in lactose-intolerant diets is not diminished. However, low lactose yoghurts may be prepared with the addition of non-dairy sugars: glucose and fructose. As a result, the lactic acid fermentation can use one or both added sugars depending on LAB strains; the residual lactose can be easily reduced with comparison to normal yoghurts (Deeth and Tamime 1981).

The lactic acid fermentation does not release lactic acid only. Galactose is found in yoghurt but related amounts appear negligible. The fermentative capacity of LAB microflora towards this sugar is strictly dependent on genetic and environmental factors such as the availability of other sources of glucidic energy. On the other hand, galactose traces might represent one of the few clinical contraindications in the use of yoghurt by galactosemic patients (Tonguç and Karagözlü 2013).

After lactic acid and galactose, the presence of glucose should be discussed. This monosaccharide is metabolized rather quickly and is detectable only in trace amounts (<0.1 %) in freshly prepared yoghurts. As a consequence, glucose does not appear to be interesting in yoghurts.

Protein fractions in yoghurt foods may be subjected to proteolytic activity because of the presence of LAB microflora: actually, only 1 or 2 % of caseins are lysed with the release of amino acids and peptides in negligible quantities.

In addition, the lipolysis on triglycerides appears without notable consequences: in fact, active lipases should be produced by spreading micro-organisms, while LAB cultures do not show similar activities.

As a result, commercially available yoghurts appear to have a chemical profile with three prevailing analytes: lactic acid, galactose and glucose. With relation to trace elements and chemical compounds, the activity of the lactic microflora leads to profound modifications in the content of water-soluble vitamins. These variations are related also to heat treatments: in summary, the content of folic acid and vitamins B_1, B_6 and B_{12} is notably modified.

Probably, the position of folic acid is interesting: this chemical is present in normal milks but also rapidly synthesized by streptococci. For this reason, the molecule is two or three times higher than the initial quantity in raw milks.

A final consideration about the fermentative LAB activity should be made with reference to the development of the aroma. Flavours of yoghurts appear mainly associated with the presence of lactic acid and acetaldehyde (Beshkova et al. 1998; Ott et al. 1997): the production of the aldehyde becomes significant when pH is ranged between 4.0 and 5.0. Small amounts of acetaldehyde and other carbonyl compounds (acetone, acetoin and diacetyl) are sufficient to give the typical flavour. When speaking of homogeneous yoghurts, the fermentation takes place in the special 'ripening' consisting of cylindrical containers.

1.7 Final Processes

The excessive acidification of yoghurts may be avoided by reducing the temperature to lower values: the aim is to inhibit the activity of used LAB cultures. The rupture of the formed clot is necessary and carried out during the initial stage of cooling. Briefly, this operation determines the first rupture of the clot and a more uniform and rapid cooling of the whole yoghurt mass with beneficial effects on the inhibition of LAB cultures. Subsequently, the yoghurt has to be forced through dedicated filters or steel discs in order to complete the breakage.

The final step is the packaging process. However, the addition of useful ingredients—fruit preparations such as puree, juice or pieces—may be carried out before this stage depending on final formulations. After this step, the fluid mass is sealed and suitably packaged. Actually, the final temperature of yoghurts is about 20 °C: this thermal condition is absolutely unsuitable to ensure the proper preservation until the consumption.

Another strategy contemplates the addition of fruit preparations to the pasteurized milk mixture before the fermentation (Chee et al. 2005). Consequently, the 'intermediate' milk mixture has to remain highly consistent and should not be subjected to ruptures of the acid clot. Subsequently, yoghurt is cooled and aseptically packaged (Vetter et al. 1974). Packaged products are then placed in special rooms where fermentative processes may continue.

References

Accolas JP, Bloquel R, Didienne R, Regnier J (1977) Propriètès acidifiantes des bactèries lactiques thermophiles en relation avec la fabrication du yoghurt. Lait 57:1–23. doi:10.1051/lait:1977561-5621

Battistotti B, Bottazzi V (1998) Microbiologia e tecnologia dei latti fermentati. In: Bottazzi V (ed) I latti fermentati. Aspetti biochimici, tecnologici, probiotici e nutrizionali. Istituto Danone, Milan, pp 35–72

Benezech T, Maingonnat JF (1994) Characterization of the rheological properties of yoghurt—a review. J Food Eng 21(4):447–472. doi:10.1016/0260-8774(94)90066-3

Benjelloun H, Rochex A, Lecouturier D, Dechemi S, Lebeault JM (2005) An on-line technique for monitoring propionic acid fermentation. Appl Microbiol Biotechnol 68(3):316–332. doi:10.1007/s00253-005-1888-3

Beshkova D, Simova E, Frengova G, Simov Z (1998) Production of flavour compounds by yogurt starter cultures. J Ind Microbiol Biotechnol 20(3–4):180–186. doi:10.1038/sj.jim.2900504

Bouillanne C, Desmazeaud MJ, Landon M, Gratadoux J, Fourier M (1980) Étude de quelques caractères de souches de Streptococcus thermophilus utilisées en fabrication de yoghourt et proposition d'une méthode de classement. Lait 60(598):458–473. doi:10.1051/lait:198059826

Buckenhüskes HJ (1993) Selection criteria for lactic acid bacteria to be used as starters coltures for various food commodities. FEMS Microbiol Rev 12:253–272. doi:10.1016/0168-6445(93)90067-J

Campbell-Platt G (1995) Fermented meats—a world perspective. In: Campbell-Platt G, Cook PE (eds) Fermented meats. Springer, New York, pp 39–52

Cappelli P, Vannucchi V (1990) Chimica degli alimenti. Zanichelli, Bologna, pp 143–147

Chandan RC, Kilara A (eds) (2013) Manufacturing yogurt and fermented milks, 2nd edn. Wiley, Chichester, pp 297–317

Chee CP, Gallaher JJ, Djordjevic D, Faraji H, McClements DJ, Decker EA, Hollender R, Peterson DG, Roberts RF, Coupland JN (2005) Chemical and sensory analysis of strawberry flavoured yogurt supplemented with an algae oil emulsion. J Dairy Res 72(03):311–316

Corradini C (1995) Chimica e tecnologia del latte. Tecniche Nuove, Milan, pp 235–236

Dalla Rosa M, Giroux F (2001) Osmotic treatments (OT) and problems related to the solution management. J Food Eng 49(2):223–236. doi:10.1016/S0260-8774(00)00216-8

De Lorenzi L, Pricl S, Torriano G (1995) Rheological behaviour of low-fat and full-fat stirred yoghurt. Int Dairy J 5(7):661–671. doi:10.1016/0958-6946(95)00047-7

De Noni I, Resmini A, Tirelli A (1998) Chimica e tecnologia dello yogurt. In: Bottazzi V (ed) I latti fermentati. Aspetti biochimici, tecnologici, probiotici e nutrizionali. Istituto Danone, Milan, pp 73–98

Deeth HC, Tamime AY (1981) Yogurt: nutritive and therapeutic aspects. J Food Prot 44:78–86

Dirar HA (1993) The indigenous fermented foods of the Sudan: a study in African food and nutrition. CAB International, Wallingford

Duncan SH, Barcenilla A, Stewart CS, Pryde SE, Flint HJ (2002) Acetate utilization and butyryl coenzyme A (CoA): acetate-CoA transferase in butyrate-producing bacteria from the human large intestine. Appl Env Microbiol 68(10):5186–5190. doi:10.1128/AEM.68.10.5186-5190.2002

Everett DW, McLeod RE (2005) Interactions of polysaccharide stabilisers with casein aggregates in stirred skim-milk yoghurt. Int Dairy J 15(11):1175–1183. doi:10.1016/j.idairyj.2004.12.004

Fleming HP, McFeeters RF, Daeschel MA (1985) The lactobacilli, pediococci, and leuconostocs: vegetable products. In: Gilliand SE (ed) Bacterial starter cultures of foods. CRC Press Inc, Boca Raton, pp 97–118

Flynn S, van Sinderen D, Thornton GM, Holo H, Nes IF, Collins JK (2002) Characterization of the genetic locus responsible for the production of ABP-118, a novel bacteriocin produced by the probiotic bacterium *Lactobacillus salivarius* subsp. *salivarius* UCC118. Microbiology 148(4):973–984

Foschino R, Galli A, Perrone F, Ottogalli G (1995) Prodotti da forno ottenuti da impasti a lievitazione naturale: studio e caratterizzazione della microflora tipica. Tec Molit Int 46(5):485–502

Fox PF, Brodkorb A (2008) The casein micelle: historical aspects, current concepts and significance. Int Dairy J 18(7):677–684. doi:10.1016/j.idairyj.2008.03.002

Franz CMAP, Von Holy A (1996) Thermotolerance of meat spoilage lactic acid bacteria and their inactivation in vacuum-packaged vienna sausages. Int J Food Microbiol 29(1):59–73. doi:10.1016/0168-1605(95)00022-4

Global Industry Analysts (2010) Global yoghurt consumption to surpass $67 Billion by 2015, According to New Report by Global Industry Analysts, Inc. http://www.prweb.com/releases/2010/02/prweb3453354.htm . Accessed 08 Apr 2014

Jokovic N, Nikolic M, Begovic J, Jovcic B, Savic D, Topisirovic L (2008) A survey of the lactic acid bacteria isolated from Serbian artisanal dairy product kajmak. Int J Food Microbiol 127(3):305–311. doi:10.1016/j.ijfoodmicro.2008.07.026

Kandler O, Weiss N (1986) Regular, nonsporing gram-positive rods. In: Sneath PHA, Mair NS, Sharpe ME, Holt JG (eds) Bergey's Manual of systematic bacteriology, vol 2. Williams and Wilkins, Baltimore, pp 1208–1234

Leroy F, De Vuyst L (2004) Lactic acid bacteria as functional starter cultures for the food fermentation industry. Trends Food Sci Technol 15(2):67–78. doi:10.1016/j.tifs.2003.09.004

Leroy F, Verluyten J, De Vuyst L (2006) Functional meat starter cultures for improved sausage fermentation. Int J Food Microbiol 106(3):270–285. doi:10.1016/j.ijfoodmicro.2005.06.027

Lindgren SE, Dobrogosz WJ (1990) Antagonistic activities of lactic acid bacteria in food and feed fermentations. FEMS Microbiol Lett 87(1–2):149–164. doi:10.1111/j.1574-6968.1990.tb04885.x

Liu SN, Han Y, Zhou ZJ (2011) Lactic acid bacteria in traditional fermented Chinese foods. Food Res Int 44(3):643–651. doi:10.1016/j.foodres.2010.12.034

Lucey JA, Fox PF (1993) Importance of calcium and phosphate in cheese manufacture: a review. J Dairy Sci 76(6):1714–1724. doi:10.3168/jds.S0022-0302(93)77504-9

Lyck S, Nilsson LE, Tamime AY (2006) Miscellaneous fermented milk products. In: Tamime AY (ed) Fermented milks. Blackwell Science Ltd, Oxford, pp 217–236

Max B, Salgado JM, Rodríguez N, Cortés S, Converti A, Domínguez JM (2010) Biotechnological production of citric acid. Braz J Microbiol 41(4):862–875. doi:10.1590/S1517-83822010000400005

McKinley MC (2005) The nutrition and health benefits of yoghurt. Int J Dairy Technol 58(1):1–12. doi:10.1111/j.1471-0307.2005.00180.x

McMahon DJ, Oommen BS (2013) Casein micelle structure, functions, and interactions. In: McSweeney PLH, Fox PF (eds) Advanced dairy chemistry: volume 1A: proteins: basic aspects, 4th edn. Springer, New York, pp 185–209. doi: 10.1007/978-1-4614-4714-6_6

McSweeney PL (2007) How does the composition of milk affect syneresis? In: McSweeney PL (ed) Cheese problems solved. Woodhead Publishing Limited, Abington, p 74

Montanari G, Zambonelli C, Grazia L, Kamesheva GK, Shigaeva MK (1996) Saccharomyces unisporus as the principal alcoholic fermentation microorganism of traditional koumiss. J Dairy Res 63(02):327–331. doi:10.1017/S0022029900031836

Motarjemi Y (2002) Impact of small scale fermentation technology on food safety in developing countries. Int J Food Microbiol 75(3):213–229. doi:10.1016/S0168-1605(01)00709-7

Narayanan N, Roychoudhury PK, Srivastava A (2004) L (+) lactic acid fermentation and its product polymerization. Electron J Biotechnol 7(2):167–178

Ott A, Fay LB, Chaintreau A (1997) Determination and origin of the aroma impact compounds of yogurt flavor. J Agr Food Chem 45(3):850–858. doi:10.1021/jf960508e

Ottogalli G (2001) Atlante dei formaggi: guida a oltre 600 formaggi e latticini pro-venienti da tutto il mondo. Ulrico Hoepli, Milan

Parisi S, Laganà P, Stilo A, Micali M, Piccione D, Delia S (2009) The maximum water absorption in cheeses. Tripartite networks of absorbed molecules per nitrogen. Ind Aliment 48(491):31–41

Park CH, Okos MR, Wankat PC (1989) Acetone–butanol–ethanol (ABE) fermentation in an immobilized cell trickle bed reactor. Biotechnol Bioeng 34(1):18–29. doi:10.1002/bit.260340104

Petridis D, Dimitreli G, Chrysalidou S, Akakiadou P (2013) Optimization of the rheological and sensory properties of stirred yogurt as affected by chemical composition and heat treatment of buffalo milk. J Food Res 2(6):55–71

Pizzoferrato L, Manzi P, Vivanti V, Nicoletti I, Corradini C, Cogliandro E (1998) Maillard reaction in milk-based foods: nutritional consequences. J Food Prot 61(2):235–239

Pot B, Tsakalidou E (2009) Taxonomy and metabolism of Lactobacillus. In: Ljungh, Wadström T (eds) Lactobacillus molecular biology: from genomics to probiotics. Caister Academic Press, Norfolk, pp 3–59

Prajapati JB, Nair BM (2003) The history of fermented foods. In: Farnworth ER (ed) Handbook of fermented functional foods. CRC Press, Boca Raton, pp 1–25

Pyler EJ (1973) Baking science and technology, vol 2. Siebel Publ. Co., Chicago

Romano P, Capece A (2013) Saccharomyces cerevisiae as bakers' yeast. Encycl Biotechnol Agric Food 158:1–4. doi:10.1081/E-EBAF-120047305

Roukas T (1991) Production of citric acid from beet molasses by immobilized cells of *Aspergillus niger*. J Food Sci 56(3):878–880. doi:10.1111/j.1365-2621.1991.tb05409.x

Ruegg PL (2005) Relationship between bulk tank milk somatic cell count and antibiotic residues. In: Proceedings of the 44th annual meeting National Mastitis Council, Orlando, 16–19 Jan 2005, pp 28–35

Savini M, Cecchini C, Verdenelli MC, Silvi S, Orpianesi C, Cresci A (2010) Pilot-scale production and viability analysis of freeze-dried probiotic bacteria using different protective agents. Nutrients 2(3):330–339. doi:10.3390/nu2030330

Schleifer KH (1986) Gram-positive cocci. In: Sneath PHA, Mair NS, Sharpe ME, Holt JG (eds) Bergey's Manual of systematic bacteriology, vol 2. Williams and Wilkins, Baltimore, pp 999–1002

Simova E, Beshkova D, Angelov A, Hristozova T, Frengova G, Spasov Z (2002) Lactic acid bacteria and yeasts in kefir grains and kefir made from them. J Ind Microbiol Biotechnol 28(1):1–6. doi:10.1038/sj/jim/7000186

Simpson BK, Nollet LM, Toldrá F, Benjakul S, Paliyath G, Hui YH (eds) (2012) Food biochemistry and food processing. Blackwell Publishing Professional, Ames

Stiles ME, Holzapfel WH (1997) Lactic acid bacteria of foods and their current taxonomy. Int J Food Microbiol 36(1):1–29. doi:10.1016/S0168-1605(96)01233-0

Tamime AY (2002) Fermented milks: a historical food with modern applications—a review. Eur J Clin Nutr 56(suppl 4):S2–S15

Tamime AY, Robinson RK (1999) Biochemistry of fermentation. In: Tamime AY, Robinson RK (eds) Yoghurt science and technology, 2nd edn. CRC Press, Cambridge

Tamime AY, Robinson RK (2007) Tamime and Robinson's yoghurt: science and technology. Woodhead Publishing Limited, Abington

Tonguç IE, Karagözlü C (2013) Galactosemia, diet and dairy products: a review. Agro FOOD Industry Hi Tech 24(6) http://www.teknoscienze.com/Articles/Agro-FOOD-INDUSTRY-hi-tech-Galactosemia-diet-and-dairy-products-a-review.aspx . Accessed 08 Apr 2014

Trejo R (2012) Casein micelles from bovine milk: native structure, interactions, and practical applications of their structural modification. Dissertation, University of Tennessee

Tuinier R, De Kruif CG (2002) Stability of casein micelles in milk. J Chem Phys 117(3):1290–1295. doi:10.1063/1.1484379

Uricanu VI, Duits MHG, Mellema J (2004) Hierarchical networks of casein proteins: an elasticity study based on atomic force microscopy. Langmuir 20(12):5079–5090. doi:10.1021/la0363736

Vetter W, Mueller H, Geissman T (1974) Preparation of fruit compositions. US Patent 3,846,570, 11 May 1974

Volonterio Galli A (2005) Microbiologia degli Alimenti. Casa Editrice Ambrosiana, Milan
Woolford MK (1984) The silage fermentation. Marcel Dekker Inc, New York, pp 247–250
Zhang Y, Zhang L, Du M, Yi H, Guo C, Tuo Y, Han X, Li J, Zhang L, Yang L (2011) Antimicrobial activity against *Shigella sonnei* and probiotic properties of wild lactobacilli from fermented food. Microbiol Res 167(1):27–31. doi:10.1016/j.micres.2011.02.006
Zourari A, Accolas JP, Desmazeaud MJ (1992) Metabolism and biochemical characteristics of yogurt bacteria. A review. Le Lait 72:1–34. doi:10.1051/lait:199211

Chapter 2
The Yoghurt: Chemical and Technological Profiles

Abstract The complex of fermentative reactions that are normally used in the industry of yoghurts shows extraordinary performances with relation to produced amounts and the qualitative composition of organic acids. The usual fermentative pathway in the yoghurt manufacture is coincident with the common homolactic fermentation. The synergic action of selected streptococci and lactobacilli in the fermentation of raw milks determines the increased production of lactic acid, acetaldehyde and polysaccharides. Other interesting variations of chemical profiles in yoghurts can be observed and explained with relation to the qualitative and quantitative distribution of different vitamins, benzoic and orotic acids, bacteriocins, enzymes, peptides and amino acids.

Keywords Acetaldehyde · Acetoin · Acetone · Bacteriocins · Diacetyl · Folic acid · Homolactic fermentation · Lactic acid · Pantothenic acid

List of Abbreviations

CO_2 Carbon dioxide
CFU Colony-forming unit
LAB Lactic acid bacterium
LDB *Lactobacillus delbrueckii* subsp. *bulgaricus*
MW Molecular weight
ST *Streptococcus thermophilus*

E. Baglio, *Chemistry and Technology of Yoghurt Fermentation*, SpringerBriefs in Chemistry of Foods, DOI: 10.1007/978-3-319-07377-4_2, © The Author(s) 2014

2.1 The Yoghurt: Biochemical Variations

The complex of fermentative reactions that are normally used in the industry of yoghurts shows extraordinary performances with relation to produced amounts and the qualitative composition of organic acids. In fact, the usual fermentative pathway in the yoghurt manufacture is coincident with the common homolactic fermentation (Sect. 1.1). However, two distinct and important features have to be mentioned:

- The fermentative pathway is carried out by two different bacteria: *Lactobacillus delbruekii* subsp. *bulgaricus* (LDB) and *Streptococcus thermophilus* (ST)
- Above-mentioned lactic acid bacteria (LAB) are able to produce notable amounts of organic acids by converting the same substrate (lactose) without competition.

On the contrary, LDB and ST can act synergically. Differently from other opportunistic associations, the synergic interaction is extremely efficient because of the increased production of lactic acid, acetaldehyde and polysaccharides (Sect. 1.4).

The synergistic effect between above-mentioned LAB may be easily explained. ST is stimulated by the bioavailability of free amino acids and peptides in culture media. On the other side, LDB is able to attack and proteolyze milk proteins; consequently, needed amino acids and peptides can be easily found if lactobacilli can spread freely in milks.

In general, it can be affirmed that five amino acids—valine, glycine, histidine, leucine and isoleucine—and short peptides are preferred sources of nitrogen for ST. With reference to LDB, the microbial growth is stimulated by active compounds such as carbon dioxide and formic acid. It has to be highlighted that similar chemicals are massively produced by ST. As a clear result, the one bacterial species seems to act as a biochemical 'growth engine' for the other life forms: in other words, the synergic association can be considered such as a 'binary feedback loop'. This situation has been repeatedly observed in different experiments with or without the addition of specified prebiotics (Oliveira et al. 2009; De Souza Oliveira et al. 2011).

It has been also noted that milk pre-treatments can reduce carbon dioxide to low values: this effect does not favour the rapid development of LDB because this bacterium would need 30 mg/Kg at least of carbon dioxide (CO_2). However, the release of CO_2 can be enhanced by decarboxylation of urea: this reaction, carried out by ST in yoghurts, has been reported to increase the initial level of CO_2 from 10 mg/kg at pH = 6.42 until more than 150 mg/Kg after only 60 min of incubation; pH may reach 5.7–5.8 at least. In addition, the higher the incubation period, the lower the pH and the higher the CO_2 release (Battistotti and Bottazzi 1998).

As a consequence, favourable conditions for the development of LDB can be obtained in milky mixtures (Sect. 1.5) after 37–40 min at 44.5 °C, in spite of the deficiency of CO_2 in original raw milks.

In addition, produced CO_2 can reach excessive values after 60 min. In fact, ST strains can easily hydrolyze dissolved urea in milk mixtures with the final conversion in CO_2 and ammonia, although the use of mutant streptococci without this ureasic activity has been recently reported (Corrieu et al. 2005).

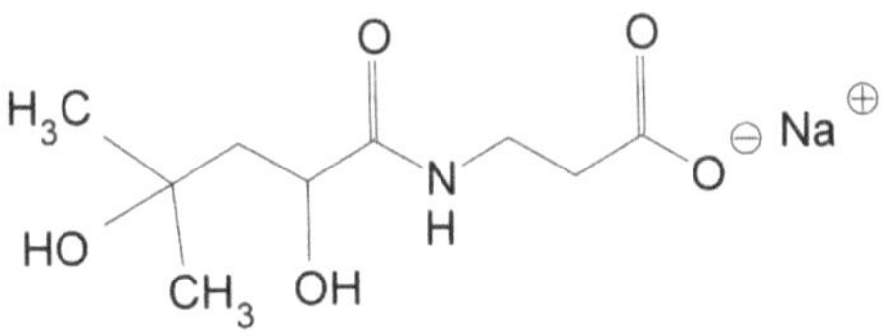

Fig. 2.1 Chemical structure of pantothenic acid, also named vitamin B_5, molecular formula: $C_9H_{16}NNaO_5$, MW: 241.216797 g mol^{-1}. BKchem version 0.13.0, 2009 (http://bkchem.zirael.org/index.html) has been used for drawing this structure

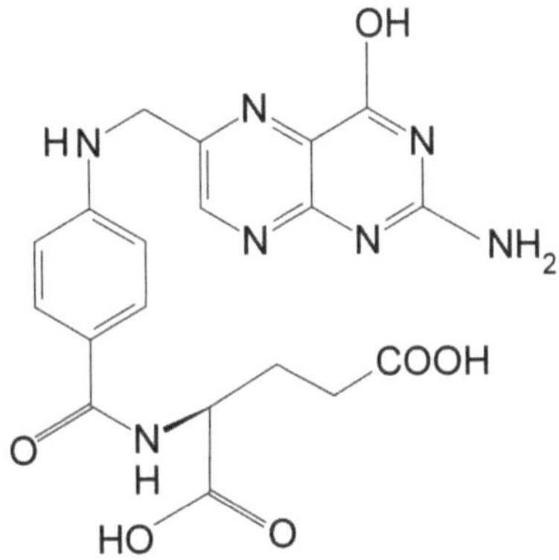

Fig. 2.2 Chemical structure of folic acid, also named vitamin B_9, molecular formula: $C_{19}H_{19}N_7O_6$, MW: 441.397491 g mol^{-1}. BKchem version 0.13.0, 2009 (http://bkchem.zirael.org/index.html) has been used for drawing this structure

Other interesting biochemical variations can occur with relation to the conversion of lactose in lactic acid—the main, declared and expected result of the homolactic fermentation—and other monosaccharides (Sect. 1.5). The amount of lactose in fresh milks is reported between 4.8 and 5.1 %; this content may be approximately 6 % when speaking of raw milk for yoghurt purposes. After evaporation and fermentation (Sect. 1.5), the chemical profile of detectable sugars should correspond to the following list: total carbohydrates, 4.8–5.2 %; unfermented lactose, 3.8–4.0 %; galactose, 1.0–1.2 %; glucose, negligible traces (Goodenough and Kleyn 1976).

LAB, homolactic fermentation and concomitant but 'extraneous' chemical reactions have other important effects on chemical profiles of vitamins, benzoic and orotic acids, bacteriocins, enzymes, peptides and amino acids.

It has been reported (Sect. 1.6) that several B vitamins—pantothenic acid, biotin, etc.—can decrease during the whole process of yoghurt productions. On the other side, the demolition of vitamins by microbial activity is not absolute but selective. For example, pantothenic acid—molecular formula: $C_9H_{16}NNaO_5$, molecular weight (MW): 241.216797 g mol^{-1}, Fig. 2.1—can diminish in yoghurts after three hours: 20 % of the original content in milks (Battistotti and Bottazzi 1998). Vitamin B_{12} can also decrease in the same condition: the measured diminution appears approximately 16.9 %. On the other side, folic acid—molecular formula: $C_{19}H_{19}N_7O_6$, MW: 441.397491 g mol^{-1}, Fig. 2.2—appears to increase. The final amount of this chemical substance—3,805 μg/100 g—is 10.3 times higher than the original content in raw milks (Battistotti and Bottazzi 1998). Other studies indicate 80 μg/L in comparison with the original content—40 μg/L—in the original milk (Chandan and Kilara 2013). Another vitamin, niacin, appears to slightly increase during time: +8.0 %.

The above-shown chemical profile of vitamins may be explained as follows: both LAB culture components require bioavailable pantothenic acid (Chandan and Kilara 2013). Additionally, LDB needs also a considerable amount of folic acid, but this molecule is abundantly synthesized by ST.

Actually, chemical profiles may always change during the storage of yoghurts. For example, the maximum production of folic acid is obtained at 42 °C during the incubation period. Subsequently, this molecule is reported to loss approximately 30 %, while vitamin B_{12} continues to decrease: related contents appear undetectable after 5 days of storage (Chandan and Kilara 2013).

It should be noted also that there is a close relation between LAB and acetic acid bacteria with reference to the synthesis of vitamin B_{12} in acid–alcoholic drinks such as *kefir*. In particular, the detection of this vitamin shows increased values when LAB and acetic acid bacteria are associated with the fluid food medium (Otles and Cagindi 2003).

With relation to organic acids, two of those normally present in the milk undergo significant transformations. Citric and uric acids do not seem normally metabolized by LAB-associated cultures. On the other side, orotic acid is reduced to 4.6 mg/100 mg ppm from the original amount of 8.2 mg/100 mg in the original milk, depending on productive conditions (Okonkwo and Kinsella 1969). Benzoic acid is present in traces in milk for yoghurt purposes: after the production, it can be found at 15–30 ppm and is associated with the metabolism of LAB cultures (Otles and Cagindi 2003). The precursor of benzoic acid, hippuric acid, can be converted depending on the activity of certain LAB strains (Tamime and Robinson 2007); in yoghurt, benzoic acid may be found between 35 and 60 ppm. Detectable contents appear approximately 30 ppm when LDB and *Lactobacillus acidophilus* are associated, while the same bacteriostatic substance is reported to be between 10 and 35 ppm when mesophilic cultures of *Lactococcus lactis* subsp. *lactis* are used (Battistotti and Bottazzi 1998).

The synthesis of above-shown acids is always considered with the concomitant production of substances with antibacterial activities: organic acids or hydrogen peroxide. With exclusive reference to yoghurts, the 'bulgarican' bacteriocin is synthesized by LDB. This substance is thermostable, active at low pH and is reported to show a wide spectrum of action towards both Gram-positive and Gram-negative micro-organisms (Otles and Cagindi 2003). ST can synthesize 'thermophilin 13', a peptide with MW of approximately 4,000 g mol^{-1}: it is thermostable and active at different pH values (Marciset et al. 1997). This LAB species produces also active bacteriocins towards moulds such as *Aspergillus* and *Rhizopus* genera.

The accumulation of beta-galactosidase, a significant probiotic, in fermented milks has to be remembered (Friend et al. 1983). The same thing can be affirmed when speaking of proteases and peptidases of microbial origin: the release of peptides and amino acids by hydrolysis is often correlated with nutritional and physiological benefits on the human health (Beshkova and Frengova 2012; Vasilijevic and Shah 2007).

On the basis of mentioned chemical profiles, it can be highlighted that the microbial metabolism of LAB cultures in yoghurts and other fermented milks determines the synthesis of following chemicals:

1. Compounds derived from the hydrolysis and the metabolism of lactose: CO_2, lactic acid, acetic acid, ethyl alcohol, formic acid, succinic acid, acetone, diacetyl, acetoin, galactose and polysaccharides
2. Compounds derived from the hydrolysis of proteins: peptides, amino acids, acetic aldehyde
3. Compounds derived from the utilization of urea: ammonia, formic acid, CO_2
4. Compounds derived from the use of organic acids: benzoic acid from hippuric acid.

Most detectable variations concerns:

- Modifications in the chemical profile of vitamins: increase in folic acid and niacin, decrease of pantothenic acid and B_{12}
- Changes in the amount of detectable nucleotides: increase of adenosine monophosphate, uridin monophosphate, guanine monophosphate and adenine dinucleotide
- Variations in the distribution of minerals: increase in ionic forms, destabilization of calcium caseinate–phosphate complexes
- Synthesis of proteins with recognized antibacterial activity: various bacteriocins
- Accumulation of enzymes: beta-galactosidase, proteases and peptidases.

Naturally, the strong increase of LAB microbial population should be also remembered: the presence of hundreds of millions per gram of LAB and other probiotic micro-organisms can notably determine the protection of the produced yoghurt (and intermediate milky mixtures during the fermentative process) against undesired life forms.

2.2 Compositional Features of Yoghurts

The qualitative and quantitative profiles of nutrients in the milk mixture are partially modified by fermentative processes: negligible modifications may be observed during storage at 4 °C because the enzymatic activity of the active microflora is slowed down or even inhibited. Actually, only a significant reduction of vitamin B_{12} and pantothenic acid amounts—60 and 30 %, respectively—and a negligible decrease of lactose and galactose have been observed. By a general viewpoint, it can be affirmed that yoghurts show a compositional situation with more abundant nutrients in comparison with the original milk, excluding lactose. Anyway, the addition of milk proteins is crucial.

Table 2.1 shows the approximate composition of commercial yoghurts in the modern industry and original data for several milks. The composition of dessert or enriched yoghurts is determined also by the nature and the quantity of non-milk ingredients. For example, glucose is particularly variable—0.0 % in white products, 0.7–3.2 % in fruit yoghurts—depending on the type of product and the peculiar manufacturer. The composition of carbohydrates concerns (De Noni et al. 1998):

Table 2.1 Average composition and energetic percentage value of cow's milk and some types of yoghurt (Agroscope Composition 2007; Agricultural Research Service 2013)

	Milk		Yoghurt		
	Full cream	Skimmed	Full cream white	Skimmed white	Full cream fruit
Dry residual, %	12.5	9.5	15	14	22
Protein, %	3.3	3.4	4.3	6.7	4.3
Fat, %	3.4	0.2	4.1	0.2	4.5
Total sugar, %	4.8	5.0	4.6	4.9	15.7
Lactic acid, %	0.003	0.003	0.5	0.5	0.5
Energetic value, kcal/100 g	62	35	66	47	102

- Glucose
- Lactose (2.6–3.6 % in white products, 2.6–3.8 % in fruit yoghurts)
- Added sugars (sucrose and fructose: up to 8.9 and 6.7 %, respectively, in sugary white products)
- Lactulose, a disaccharide consisting of galactose and fructose, in form of concentrated syrups or high-purity crystals, obtained by means of the epimerization of lactose during the pasteurization process (Andrews 1984).

Normally, lactulose—also named 4-O-β-D-galactopyranosyl-D-fructose—is found between 0.02 and 0.07 % in sterilized milks depending on applied time/temperature cycles. On the other hand, lactulose is well known as a good sweetener: different synthetic processes are available at present with relation to the production of this substance (Aider and Halleux 2007). Theoretically, lactulose should be obtained by means of the isomerization reaction of lactose in alkaline media. One of favourite mechanisms, the reaction of Lobry de Bruyn–van Ekenstein, converts the glucose part of the original lactose in fructose by means of the formation of the enolic intermediate shape of lactose and epilactose in alkaline media such as calcium hydroxide and potassium hydroxide (Aider and Halleux 2007). However, several methods have been developed with the aim of enhancing the production: different catalyzers and technological solutions, such as the purification and the demineralization by anion and cation exchange resins have been recently proposed (Aider and Halleux 2007; Carobbi and Innocenti 1990; Carobbi et al. 1985).

The chemical profile of organic acids should be studied and discussed. With relation to lactic acid, the related content in all commercially available yoghurts appears greater than 0.8 %. This amount corresponds to the minimum quantity for this type of fermented milk. Normally, the maximum content should not exceed 1.3 %: otherwise, products might exhibit particularly acidic tastes with some risk of consumeristic unacceptability. Other milk-related constituents may be considered important on the nutritional level even if their amount is detected in limited quantities.

For instance, formic acid—produced by ST—is recognized capable of stimulating the multiplication of lactobacilli. Small amounts of this acid—less than 40 mg/kg—are also produced after the caramelization of lactose because of thermal treatments such as pasteurization (De Noni et al. 1998; Kroh 1994).

Flavours can be very important when speaking of yoghurts and the correlated commercial success. Basically, aroma is affected by the compulsory presence of minor constituents without fermentative origin: actually, these chemicals seems be produced by reactions promoted by thermal processes in the same way of formic acid.

Generally, flavours are determined by the qualitative and quantitative presence of carbonylic acids. These substances may be obtained from fats lactones, and their concentration is crucial with concern to more or less pleasant and intense tastes (Cheng 2010; De Noni et al. 1998; Gaafar 1992; Laye et al. 1993):

- Acetaldehyde (2–40 mg/kg)
- Acetoin (2–6 mg/kg)
- Acetone (1–4 mg/kg)
- Diacetyl (0.5–1.0 mg/kg).

Finally, other minor substances may be cited: total free fatty acids, octanoic acid, 9-decenoic acid, tetradecanoic acid, octanol, ethyl acetate, etc., depending on the peculiar aroma and the desired result. On the other hand, off-flavours with unacceptable or questionable results appear to be correlated with the presence of tyrosine, acetic acid, 3-methyl butanol, ethyl acetate, etc. (Cheng 2010).

In fact, the process of thermal caramelization and the so-called Maillard reaction—in which lactose is involved—can be cause of the production of these compounds. In addition, the overall chemical composition remains unmodified during storage while microbiological profiles appear be subjected to major modifications. Actually, the total number of LAB in yoghurts is comprised between 10^8 and 10^9 colony-forming units (CFU)/g during the production. Subsequently, a notable reduction is observed after 30 days at 4 °C until a 'final' 10^7 CFU/g microbial count, although some discordance may be found in the scientific literature depending on storage temperatures and other factors (Birollo et al. 2000).

Higher storage temperatures cause generally more rapid decreases of microbial counts until to 5×10^6 CFU/g before of the end of shelf life. It may be supposed that this level is critical for LAB counts in yoghurts. Anyway, the presence of relatively high numbers of vital LAB cultures depends also on the respect of low storage conditions during the declared shelf life. The same thing can be easily affirmed when speaking of healthy properties and sensory—or chemical—profiles of yoghurts.

References

Agricultural Research Service (2013) USDA National Nutrient Database for Standard Reference, Release 26, Nutrient Data Laboratory Home Page. http://www.ars.usda.gov/ba/bhnrc/ndl. Accessed 8 Apr 2014

Agroscope Composition (2007) FOAG Swiss Federal Office for Agriculture. http://www.agroscope.admin.ch/joghurt-sauermilchprodukte/01368/index.html?lang=en. Accessed 8 Apr 2014

Aider M, Halleux DD (2007) Isomerization of lactose and lactulose production: review. Trends Food Sci Technol 18(7):356–364. doi:10.1016/j.tifs.2007.03.005

Andrews GR (1984) Distinguishing pasteurized, UHT and sterilized milks by their lactulose content. Int J Dairy Technol 37(3):92–95. doi:10.1111/j.1471-0307.1984.tb00498.x

Battistotti B, Bottazzi V (1998) Microbiologia e tecnologia dei latti fermentati. In: Bottazzi V (ed) I latti fermentati. Aspetti biochimici, tecnologici, probiotici e nutrizionali. Istituto Danone, Milan, pp 35–72

Beshkova D, Frengova G (2012) Bacteriocins from lactic acid bacteria: microorganisms of potential biotechnological importance for the dairy industry. Eng Life Sci 12(4):419–432. doi:10.1002/elsc.201100127

Birollo GA, Reinheimer JA, Vinderola CG (2000) Viability of lactic acid microflora in different types of yoghurt. Food Res Int 33(9):799–805. doi:10.1016/S0963-9969(00)00101-0

Carobbi R, Innocenti F (1990) Process for preparing lactulose from lactose by epimerization with sodium aluminate. US Patent 4,957,564, 18 Sep 1990

Carobbi R, Franci V, Miletti S (1985) Process for preparing lactulose from lactose, in the form of a syrup or a crystalline product. US Patent 4,536,221, 20 Aug 1985

Chandan RC, Kilara A (eds) (2013) Manufacturing yogurt and fermented milks, 2nd edn. Wiley, Chichester, pp 297–317

Cheng H (2010) Volatile flavor compounds in yogurt: a review. Crit Rev Food Sci Nutr 50(10):938–950. doi:10.1080/10408390903044081

Corrieu G, Monnet C, Sepulchre AM (2005) Use of strains of Streptococcus thermophilus which are incapable of hydrolyzing urea in dairy products. US Patent 6,962,721, 8 Nov 2005

De Noni I, Resmini A, Tirelli A (1998) Chimica e tecnologia dello yogurt. In: Bottazzi V (ed) I latti fermentati. Aspetti biochimici, tecnologici, probiotici e nutrizionali. Istituto Danone, Milan, pp 73–98

De Souza Oliveira RP, Perego P, De Oliveira MN, Converti A (2011) Effect of inulin as a prebiotic to improve growth and counts of a probiotic cocktail in fermented skim milk. LWT-Food Sci Technol 44(2):520–523. doi:10.1016/j.lwt.2010.08.024

Friend BA, Fiedler JM, Shahani KM (1983) Influence of culture selection on the flavor, antimicrobial activity, beta-galactosidase and B-vitamins of yoghurt. Milchwissenschaft 38:133–136

Gaafar AM (1992) Volatile flavour compounds of yoghurt. Int J Food Sci Technol 27(1):87–91. doi:10.1111/j.1365-2621.1992.tb01183.x

Goodenough ER, Kleyn DH (1976) Qualitative and quantitative changes in carbohydrates during the manufacture of yogurt. J Dairy Sci 59(1):45–47. doi:10.3168/jds.S0022-0302(76)84153-7

Kroh LW (1994) Caramelisation in food and beverages. Food Chem 51(4):373–379. doi:10.1016/0308-8146(94)90188-0

Laye I, Karleskind D, Morr CV (1993) Chemical, microbiological and sensory properties of plain nonfat yogurt. J Food Sci 58(5):991–995. doi:10.1111/j.1365-2621.1993.tb06096.x

Marciset O, Jeronimus-Stratingh MC, Mollet B, Poolman B (1997) Thermophilin 13, a nontypical antilisterial poration complex bacteriocin, that functions without a receptor. J Biol Chem 272(22):14277–14284

Okonkwo P, Kinsella JE (1969) Orotic acid in yoghurt. J Dairy Sci 52(11):1861–1862. doi:10.3168/jds.S0022-0302(69)86856-6

Oliveira RPDS, Perego P, Converti A, De Oliveira MN (2009) Effect of inulin on growth and acidification performance of different probiotic bacteria in co-cultures and mixed culture with Streptococcus thermophilus. J Food Eng 91(1):133–139. doi:10.1016/j.jfoodeng.2008.08.013

Otles S, Cagindi O (2003) Kefir: a probiotic dairy-composition, nutritional and therapeutic aspects. Pak J Nutr 2(2):54–59

Tamime AY, Robinson RK (2007) Tamime and Robinson's yoghurt: science and technology. Woodhead Publishing Limited, Abington

Vasilijevic T, Shah NP (2007) Fermented milk: health benefits beyond probiotic effect. In: Hui YH, Chandan RC, Clark S, Cross NA, Dobbs JC, Hurst WJ, Nollet LM, Shimoni E, Sinha N, Smith EB, Surapat S, Titchenal A, Toldrá F (eds) Handbook of food products manufacturing: health, meat, milk, poultry, seafood, and vegetables, Vol. 2. John Wiley & Sons, Hoboken, pp. 99–115

Chapter 3
The Industry of Yoghurt: Formulations and Food Additives

Abstract Notable modifications have changed the industry of dairy products in recent years. The preference of a large part of consumers for yoghurts may be partially explained with the composition of initial raw materials. On the other hand, marketing strategies and the increasing attention to new dietary advices have progressively enhanced the consumption of new products. As a consequence, the modification of yoghurt formulations should be observed in the broader ambit of the milk industry. Anyway, the commercial classification of yoghurt types may be extended to a number of different sub-classes depending on the peculiar request of the final consumer or user. The formulation of yoghurts influences a small group of product features such as texture and syneresis. For these reasons, the modern industry can decide for the use of selected chemical additives with the aim of producing ameliorated versions of the same product or sub-typology. This argument is discussed here with reference to four classes of food additives.

Keywords Food additive · Food classification · Food design · Rheology · Shelf life · Yoghurt

List of Abbreviations

G	1,4-linked α-L-guluronic acid
M	1,4-linked β-D-mannuronic acid
CAS	Chemical Abstracts Service
CA	Codex Alimentarius Commission
EFSA	European Food Safety Authority
FCS	Food Category System

E. Baglio, *Chemistry and Technology of Yoghurt Fermentation*, SpringerBriefs in Chemistry of Foods, DOI: 10.1007/978-3-319-07377-4_3, © The Author(s) 2014

FP Food producer
MW Molecular weight
KOH Potassium hydroxide

3.1 The Yoghurt in the Modern Industry: An Overview

Notable modifications have profoundly changed the industry of dairy products in recent years. Many different factors may be taken into account when considering the substantial evolution of yoghurt productions (Kaminarides et al. 2007):

- Composition of milk (ovine and bovine types)
- New consumers' needs in relation to dietary prescriptions
- Marketing choices
- New requests by food processors (users of yoghurt and other milk products)
- The progressive availability of typical and regional yoghurt recipes in many markets.

This list is not exhaustive. However, the evolution of yoghurt products appears to be influenced by above-mentioned factors.

In detail, the preference of a large part of consumers in selected countries may be partially explained with the composition of initial raw materials (Alichanidis and Polychroniadou 1996; Anifantakis 1986; Kaminarides et al. 2007). Substantially, the yoghurt made from sheep milk seems to contain increased amounts of triglycerides, monounsaturated fatty acids and linolenic acid if compared with yoghurts from bovine milk. In addition, the bioavailability of selected mineral elements—calcium, iron, zinc, etc.—and vitamins appears higher in the first mentioned type. Finally, the proportion of proteins and essential amino acids is favourable when speaking of 'ovine' yoghurts.

As a result, the preference of many consumers may be reasonably explained; in addition, organoleptic features of yoghurts from ovine milk appear to be very pleasant (Kaminarides et al. 2007; Kurmann 1986). Aroma and flavour properties of yoghurts are mainly influenced by the presence of peculiar non-volatile or volatile acids and carbonyl compounds (Kaminarides et al. 2007). This factor highlights the important role of starter cultures in the fermentation of raw materials (Tamime and Robinson 1999).

On the other side, marketing strategies and the increasing attention to new dietary advices have progressively enhanced the consumption of new products with low amounts of lipids (Kaminarides et al. 2007). These productions can be made possible when new technologies are applied with the aim of reducing the initial content of triglycerides in the initial milk. As a clear consequence, the modification of yoghurt formulations should be observed in the broader ambit of the milk industry. For example, the separation of bovine and ovine milks into economically interesting sub-products—butters, creams, whey proteins, caseins, lactose, etc.—can potentially influence notable portions of current food markets. The influence of similar productions can affect dairy and non-dairy sectors of the food production with remarkable results.

Anyway, the commercial classification of yoghurt types appears to consider three main classes at least (Benezech and Maingonnat 1994; Magenis et al. 2006):

- Set-type products
- Stirred yoghurts
- Drinking yoghurts.

Different consumers may prefer different products; consequently, the yoghurt industry has progressively defined the above-mentioned separation between similar but specific typologies on the basis of two requests:

1. Consumers' exigencies
2. Needs of food producers (FP) with or without direct correlation with the production of milk/dairy products.

For instance, many industrial snack products and cereal biscuits may contain flavoured or normal yoghurts in their formulation. Because of the different destination, yoghurt technologists may be requested to define new formulations or modified 'recipes'. As a result, the tripartite classification of yoghurt products may be extended to a number of different sub-types depending on the peculiar request of the final consumer or user.

The formulation of yoghurts influences a small group of product features. Basically, it may be assumed that texture and syneresis are two of main technological factors for yoghurt technologists (Magenis et al. 2006). On these bases, following variables have to be carefully managed when 'designing' a new yoghurt (Harwalkar and Kalab 1986):

- Original composition of raw milk (salts, ratio of caseins to whey proteins, etc.)
- Heat pre-treatment of raw milks
- Total solid amount
- Starter culture
- Process of homogenization
- Final acidity
- The possible addition of chemical compounds in the formulation.

The last point appears extremely important when speaking of modern yoghurt products. For instance, rheological properties are related to processing parameters: They can influence the final texture and the consumeristic perception of final products (Aichinger et al. 2003; Benezech and Maingonnat 1994; Kaminarides et al. 2007). However, the modern industry can decide for the use of selected chemical additives (or classes of food additives) with the aim of producing ameliorated versions of the same product or sub-typology. This argument can be extremely interesting if traditional and regional types of yoghurts are placed on the market of foods and raw materials for the production of cakes, biscuits, snacks, etc.

For instance, the difference between normal yoghurts and similar fermented products (*Dahi, Doogh, Ayran, Labneh*) by undefined starter cultures and/or goat, cow and buffalo milk might be important if the second typology of product has to be used for industrial applications (Abu-Jdayil et al. 2000; Kiani et al. 2008;

Mohameed et al. 2004; Nahar et al. 2007; Younus et al. 2002). Other important examples are the production of 'enriched' yoghurts by means of the addition of orange fibres (Sendra et al. 2010), inulin (Guggisberg et al. 2009), folic acid (Aryana 2003), etc.

For these reasons, the market of yoghurt and food products with added yoghurt should be carefully examined in relation to the possible use of selected chemicals and related effects on physicochemical features of the final product. Moreover, peculiar compounds of microbial or synthetic origin may influence sensorial properties of yoghurts. However, the classification of available yoghurts should be reviewed and discussed in detail.

3.2 The Yoghurt in the Modern Industry: A Food Classification

At present, many different classifications may be proposed worldwide in relation to food and beverage products. Food descriptions may be made by means of three different strategies on condition that every typology is correctly classified with dedicated names, information on sources (raw materials), procedures for the production, manipulation and storage of finished products, preservation and cooking methods (where applicable), declared or allowed use of food additives (Codex 1995; Schlotke et al. 2000; Truswell et al. 1991):

- Monohierarchical classification systems like Eurocode 2, the 'Food Category System' (FCS) or proprietary food grouping systems when available on a national scale
- Faceted description systems using standardized thesauri like the LanguaL system, or free text.

With reference to the reliable classification of yoghurts and yoghurt-related products, the hierarchical strategy appears workable. One of the most recognized classification systems is the FCS (Codex 1995) because of the general applicability. On the other hand, this strategy has been used with the aim of creating a useful tool for assigning food additive uses in the ambit of the Codex Alimentarius Commission (CA). On these bases, all mentioned descriptors do not represent legal product designations; in addition, the system is not intended for labelling purposes (Codex 1995).

At first sight, the CA has tacitly proposed the subdivision of yoghurts as follows (Codex 1995):

- First group: drinking (flavoured, coloured) yoghurts
- Second group: fermented or plain yoghurts, with the exclusion of the first group
- Dairy-based fruit or flavoured desserts.

Above-mentioned descriptors should be discussed in detail because of the multiplicity of possible food products and the allowed addition of peculiar chemicals. In addition, the CA clearly states that the formulation of fermented plain milks should contain only stabilizers and thickeners with reconstitution and recombination functions; on the other hand, national legislations might allow the use of different food additives for the production of all types of yoghurt products (Codex 1995, 2012).

3.2.1 Drinking Yoghurts

The first group of yoghurts is included into the 01.1.2 food category, (Codex 1995). Actually, this group concerns all types of dairy-based drinks, including flavoured and/or fermented products with the exception of mixes for cocoa (cocoa–sugar mixtures).

The proposed description includes strawberry-flavoured yoghurt drinks and lactic acid bacteria drinks with the contemporary presence of chocolate malt drinks, *lassi* foods, etc. (Codex 1995). Actually, this situation depends on the possible use of a specified food additive for yoghurts and other similar drinking products at the same time. The above-shown approach may help readers to consider the design of industrial products: the final performance (enhanced colours and/or flavours, ameliorated viscosity, etc.) may justify the use of the same chemical substance in different products.

According to the proposed description, only flavoured or aromatized yoghurts (and their mixes) can be considered in this group. As a result, the proposed classification makes a clear discrimination between 'flavoured' products and mixes on the one side and 'plain' yoghurts on the other side (Sect. 3.2.2). This approach is specifically based on the declared use of food additives (Codex 1995); consequently, it may be inferred that the industrial classification of yoghurts should be based on the designed and declared addition of chemicals and/or non-dairy ingredients.

3.2.2 Fermented (Plain) Yoghurts

'Normal' or plain yoghurts are considered in the food category 01.2.1: fermented and renneted milk products, with the exclusion of dairy-based drinks (Codex 1995).

Basically, these products are obtained from skimmed, partially skimmed, low-fat and whole milks. Apparently, the origin of milk is not important with reference to this description. Consequently, the abundance of caseins (if compared with whey proteins) or triglycerides (e.g. amount of monounsaturated fatty acids and linoleic acid) is not a discriminatory function in this ambit, while the same argument appears extremely important in the research field (Sect. 3.1).

Moreover, plain yoghurts are further subdivided depending on the possible heat treatment after fermentation: in detail, not heat-treated yoghurts have not been sterilized or pasteurized (category 01.2.1.1), while thermally treated products fall in the category 0.1.2.1.2.

Once more, the proposed classification is mainly based on the possible use of one or more food additives for the peculiar product; consequently, the aim of this description is not linked to compositional properties of raw milks, fermented intermediates or final products. In fact, the use of goat, cow or sheep milk is not a discriminator for the technical definition of 'plain' yoghurt: basically, the main difference appears correlated with the sensorial detection of easily visible features such as colour, aroma or texture (all mentioned foods are defined 'fluid and non-fluid' products).

3.2.3 Dairy-Based Desserts

The third group of yoghurt products is included into the 01.7 food category (Codex 1995). Actually, this category concerns all types of dairy-based desserts, including also mixed products, frozen dairy specialties and dairy-based fillings.

The CA states clearly that flavoured yoghurts are obtained by means of the fermentation of milk and milk products (the mixture of similar raw materials with different origin is allowed) and the addition of flavours and special ingredients such as fruits, coffee, and cocoa. This definition may appear also good for the first group of yoghurts (Sect. 3.2.1). On the other hand, drinking products are substantially ready-to-eat dairy foods with low viscosity (Tamime and Robinson 1999), while the category of dairy-based desserts should exhibit very different rheological properties. With exclusive concern to drinking products, the reduced viscosity and textural features are obtained by dilution and addition of sweeteners, flavours and/or fruit juices.

In relation to yoghurts, another feature of dairy-based desserts is related to the possibility of heat treatment after fermentation (Codex 1995).

3.2.4 Other Yoghurt-Related Products

The proposed classification of yoghurt products may be exhaustive enough. On the other side, the industrial market shows often a number of different foods or beverages with some similarity in the structure or the name.

The above-mentioned FCS has examined the question with the definition of other yoghurt-related food products: the presence or the claimed similarity of fermented plain milks may be explicit or recalled in the list of ingredients.

A peculiar food category concerns yoghurt-coated processed nuts (Codex 1995). With reference to the 15.2 food category, the function of yoghurts should be highlighted: 'fermented milks' are considered here with a mere role of 'coating agent'. This reflection should suggest interesting consequences with reference to the formulation of related yoghurt 'coatings'.

In other terms, the yoghurt is defined 'perishable' or 'highly perishable' depending on different features including pH, acidity, and water content. After all, fermented milks are 'living' foods; consequently, a certain perishability should be always expected. However, the possible use as 'food coating' implies that the compositional structure of the initial yoghurt should be different enough from the normal formula: yoghurts should be resistant against oxidability at least before coating. In addition, similar raw materials should maintain a constant viscosity in spite of the demonstrated and irreversible tyxotropy (Penna et al. 2001).

As a result, selected food additives should be added to the original formulation. For instance, sodium caseinates (Collet et al. 2004), pectin (Basak and Ramaswamy 1994) and xanthan gum (Zhengtao et al. 2009) might be used in relation to rheological properties because the produced yoghurt may not be successfully applied on nut surfaces without constant viscosity. In addition, storage temperatures have to be maintained always constant. The management of similar 'coatings' may impose a profound reformulation of original yoghurt recipes.

Similar problems may be discussed when speaking of biscuit commercial preparations and similar bakery products (Zanjani et al. 2012). The use of selected chemicals is crucial with reference to rheological properties of food products, while lactic microflora should be a good warranty against the menace of pathogen and degrading bacteria. Once more, prepared yoghurts should remain viscous enough without phase separations before the final dispersion on cookies surfaces (Borneo et al. 2007). In addition, yoghurt powders are available for the production of similar foods (Koç et al. 2010): naturally, the problem could be correlated with the 'right' reconstitution of yoghurts.

Other situations may occur when yoghurt is used as a basic component of drinkable soya snacks (Chen et al. 2010; Kälviäinen et al. 2003). This time, rheological properties may be important because of the 'foreign' origin of soya proteins: the formulation may contain high-protein contents and low amount of carbohydrates.

The description of yoghurt products cannot be considered exhaustive in relation to the proposed FCS. In fact, following food product groups may contain—or be filled with—yoghurt (Codex 1995):

- Bakery wares, categories 07.1.1 and 07.1.6. For example, filled bread rolls and mixes
- Cakes, cookies and pies, category 07.2.2. For example, filled muffins (Altman and Landis 1995)
- Fruit salad with yoghurt dressing (Park et al. 2002)
- Yoghurt-flavoured candies (Peterson 1979).

Every single sub-class of industrial food can naturally constitute a notable challenge for food technologists and research chemists because of the necessity of producing a peculiar yoghurt formulation with two distinct properties:

1. Constant rheological features during a specified temporal limit before the use by FP and
2. Acceptability from the microbiological and chemical viewpoints when the yoghurt is associated (mixed, deposed as coating, etc.) with other ingredients of the final food or beverage.

In other words, the performance of yoghurts depends strongly on the final and declared use. Consequently, food additives may be not strictly necessary for a traditional recipe; on the other side, enhanced properties (extended shelf life, improved texture, etc.) may force FP to use food-grade chemicals with different functions.

3.3 Additives for Yoghurt and Yoghurt-Related Food Products

The science of food chemistry is inconceivable without a solid knowledge of food additives. This statement is certainly correct when speaking of the increasing 'weight' of chemicals in the production of modern yoghurts.

By a general viewpoint, food additives correspond to a defined subset of the whole group of edible raw materials for food manufacture, processing, preparation and treatments (Codex 1995). In detail, these chemicals (or groups of substances with a single definition) are not normally consumed as a food by itself; in addition, they are not usually included in the ingredient list of a peculiar food as a typical substance (Codex 1995).

The presence and the usefulness of food additives are always correlated with the necessity of obtaining foods or beverages with peculiar technological properties or sensorial advantages. The use of food additives can imply that these advantages or functional properties may not be obtained with other technologies. Anyway, the presence of food-grade chemicals does not suggest or imply nutritional advantages (Codex 1995). On the contrary, several toxicological concerns may be signalled when speaking of these substances (Maga and Tu 1995).

In other words, food additives should be used at least when (Codex 1995):

- Appreciable advantages are obtained
- The consumers' health is not damaged
- The final food does not mislead the consumer
- Technological functions or properties cannot be obtained by other means that are economically and technologically practicable.

Anyway, food additives can be a notable challenge for interested readers and professionals in many industrial sectors and in the academia (Smith 1991). On these bases, a circumscribed discussion within well-specified boundaries would be better understood. Unfortunately, the yoghurt industry can use—or be correlated with the use of—different food chemicals and additives. Consequently, the best strategy for the comprehension of yoghurt formulations should imply the discrimination of chemicals on the basis of a specific function (Smith 1991).

A useful scheme for general food products can consider the following 'voices' (Smith 1991):

- Antioxidants
- Sweeteners
- Flavourings
- Food colours
- Preservatives
- Enzymes
- Nutritive additives
- Emulsifiers
- Bulking agents
- pH control agents
- Hydrocolloids
- Antifoams and release agents
- Raising agents
- Gases.

By contrast, a possible yoghurt-oriented approach may take the following sub-categories into account (Codex 1995):

- Sweeteners
- Flavour enhancers
- Food colours
- Emulsifiers, sequestrants, stabilizers, thickeners
- Antioxidants
- Preservatives
- Stabilizers without other functions
- Acidity regulators
- Humectants
- Raising agents
- Anticaking agents.

This section is dedicated to the study of these categories and related roles in the formulation of modern yoghurts and yoghurt-related food products. Because of the remarkable number of food additives, four main categories—sweeteners, flavouring agents, food colours and thickeners—are mainly discussed here.

These additive groups can influence (a) sensorial properties and (b) technological features of the final yoghurt or yoghurt-related foods. Remaining food additive categories—antioxidants, preservatives, emulsifiers and stabilizers, acidity regulators, humectants, raising agents and anticaking agents—are not discussed in detail in spite of their remarkable importance. However, a simplified and not exhaustive list of most known or usable additives for these classes (Akpan et al. 2007; Codex 1995; Jaziri, et al. 2009; Modler and Kalab 1983; Najgebauer-Lejko 2014; Serrano et al. 1991; Tamime and Robinson 1999) is shown in Tables 3.1 and 3.2.

Table 3.1 A selection of antioxidant and preservative agents for the production of yoghurts and yoghurt-related foods (Akpan et al. 2007; Codex 1995; Jaziri, et al. 2009; Modler and Kalab 1983; Najgebauer-Lejko 2014; Serrano et al. 1991; Tamime and Robinson 1999)

Food additive class	Substance and other identifications
Antioxidants	Ascorbyl palmitate
	Ascorbyl stearate
	Propyl gallate
	Tea catechins (natural antioxidants)
	Ascorbic acid (natural antioxidants)
	β-carotene, capsanthin, capsorubin, cryptocapsin (carotenoids, natural antioxidants)
	Quercetin, luteolin, capsaicinoids (natural antioxidants)
Preservatives	Benzoic acid
	Sodium benzoate
	Potassium benzoate
	Calcium benzoate
	Ethyl para-hydroxybenzoate
	Methyl para-hydroxybenzoate
	Lauric arginate ethyl ester
	Sorbic acid
	Sodium sorbate
	Potassium sorbate
	Calcium sorbate
	Nisin (natural preservative)

Table 3.2 A selection of stabilizers, sequestrants, acidity regulators, humectants, raising and anticaking agents for the production of yoghurts and yoghurt-related foods (Akpan et al. 2007; Codex 1995; Jaziri, et al. 2009; Modler and Kalab 1983; Najgebauer-Lejko 2014; Serrano et al. 1991; Tamime and Robinson 1999)

Emulsifiers and stabilizers	Diacetyltartaric and fatty acid esters of glycerol	Sequestrant agent
	Calcium polyphosphate	Sequestrant agent
	Ammonium polyphosphate	Sequestrant agent
	Polyoxyethylene (20) sorbitan monolaurate	
	Polyoxyethylene (20) sorbitan monooleate	
	Polyoxyethylene (20) sorbitan monostearate	
	Polyoxyethylene (20) sorbitan tristearate	
	Sodium caseinate, whey proteins, caseins	
	Xanthan	Thickening agent
	Modified starches	
	Agar, alginates, carrageenan, etc. (seaweed extracts)	Gelling (thickening) agents
	Soy proteins	
Acidity regulators	Calcium polyphosphate	Humectant agent, raising agent

(continued)

Table 3.2 (continued)

Humectants	Calcium polyphosphate	Acidity regulator, raising agent
	Ammonium polyphosphate	
	Bone phosphate	Anticaking agent
Raising agents	Calcium polyphosphate	Humectant agent, acidity regulator
Anticaking agents	Bone phosphate	Humectant agent
	Sodium aluminosilicate	

3.3.1 Sweeteners

Sweeteners are normally added with the declared aim of enhancing sweet properties of the finished product. Consequently, modified foods should have a distinct taste of sugar or similar sugar-related ingredients such as honey or saccharin.

Depending on different contests, available sweeteners may be named as follows:

- Natural compounds
- Synthetic substances
- 'Caloric' compounds
- 'Non-caloric' or 'low-calorie' substances
- Nutritive compounds
- Non-nutritive substances.

Basically, sweeteners should be (O'Brien Nabors 2011):

1. Sweet as sucrose, non-cariogenic, colourless and odourless
2. Water soluble and stable when dissolved in acid or alkaline solutions or media
3. Non-toxic and metabolized without health damages or excreted without modifications.

Because of the main and declared function of sweeteners, the perceived effect and other accessory features (duration, aftertaste, etc.) are extremely important. Different factors may modify this feature (O'Brien Nabors 2011):

- The chemical concentration in the food or beverage
- The serving temperature
- pH
- Possible mixtures between two or more different sweeteners (classical examples are cyclamate and saccharin, and aspartame and acesulfame potassium)
- The sensitivity of the consumer.

Anyway, the use of sweeteners—artificial compounds above all—has been often criticized in relation to sugar intake or possible health consequences (American Dietetic Association 2004; Glinsmann et al. 1986). For these reasons, the increasing amount of alternative and 'natural' sweeteners is generally observed with great interest.

With exclusive relation to yoghurts and yoghurt-related foods and in accordance with the CA, the following list shows most interesting sweeteners (Codex 1995):

- Potassium acesulfame, additional function: flavour agent
- Alitame
- Aspartame, additional function: flavour agent
- Aspartame–acesulfame salt
- Cyclamic acid
- Calcium cyclamate
- Sodium cyclamate
- Neotame, additional function: flavour agent
- Saccharin
- Calcium saccharin
- Potassium saccharin
- Sodium saccharin
- Steviol glycosides
- Sucralose (trichlorogalactosucrose).

Every sweetener, both natural and artificial types, can have interesting and peculiar properties if compared with similar compounds. However, the basic aim of this book is to provide a simplified description of some food additives in relation to the related strategy for the production of yoghurt and yoghurt-related foods. For this reason, a single compound will be shortly discussed in relation to sweeteners and other functional classes. The interested reader is invited to consult other references for a detailed description of food additives. With concern to sweeteners, potassium acesulfame is chosen.

This molecule is a peculiar compound because two different functions are ascribed: sweetener and flavour enhancer. It is also known as E950, 6-methyl-3,4-dihydro-1,2,3-oxathiazin-4(3H)-one 2,2-dioxide potassium salt, or 'Sweet One'.

From the chemical viewpoint, the molecular formula is $C_4H_4KNO_4S$ with molecular weight (MW): 201.24 g mol^{-1} and Chemical Abstracts Service (CAS) registry number 55589-62-3 (Fig. 3.1).

It can be described as a crystalline, odourless and colourless powder, although white powders may be produced; the sweet taste is intense (Klug and von Rymon-Lipinsky 2011; Rowe et al. 2003).

Potassium acesulfame is normally obtained by acetoacetic acid *tert*-butyl ester and fluorosulphonyl isocyanate and the subsequent addition of potassium hydroxide (KOH) or by diketene and amidosulphonic acid with dehydrating compounds and the addition of KOH (Rowe et al. 2003).

With reference to stability, potassium acesulfame is storable for many years without appreciable consequences, although high storage temperatures may cause decomposition after several months (Klug and von Rymon-Lipinsky 2011; Rowe et al. 2003).

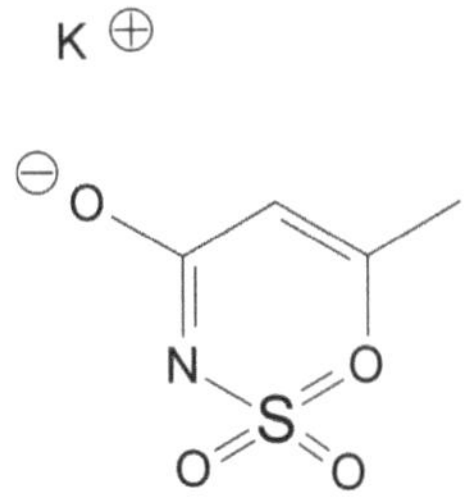

Fig. 3.1 Chemical structure of potassium acesulfame, a sweetening agent. BKchem version 0.13.0, 2009 (http://bkchem.zirael.org/index.html) has been used for drawing this structure

When dissolved in water, it remains sweet for 24 months at least. In addition, the sweetness of potassium acesulfame appears substantially unmodified after sterilization and pasteurization treatments.

Finally, the use of potassium acesulfame has been described in combination with other synthetic sweeteners such as sodium cyclamate or aspartame. However, a mixture of three different sweeteners may decrease the perceived taste if saccharin and acesulfame potassium are considered (Klug and von Rymon-Lipinsky 2011; Rowe et al. 2003; Schiffmann et al. 2003).

The strategy of the use of potassium acesulfame in the yoghurt industry considers generally mixtures with other sweeteners: this choice can be interesting when synthetic molecules are used. However, the combination of potassium acesulfame with natural sweeteners may be also useful and recommended. Several steviol glycosides like rebiana—a high-purity rebaudioside A—may be mixed with potassium acesulfame with the aim of 'mitigating' 'off' taste such as bitter and black licorice (Prakash et al. 2008).

All possible yoghurt—plain, strawberry, low fat, etc.—types can be produced with the addition of potassium acesulfame. Normally, reported evidences show a modest difference between artificial sweeteners and natural compounds with reference to the effect on the growth on micro-organisms and the viscosity of products during shelf life (Keating and White 1990).

Anyway, the best application for this type of sweetener and related mixtures appears the production of fruit-flavoured products, stirred and set-style yoghurts, in comparison with foods with separate fruit pieces (Klug and von Rymon-Lipinsky 2011). Finally, the stability of potassium acesulfame is good with reference to pasteurization and shelf life durations. Consequently, it can be assumed that this sweetener may be usable in yoghurts and other dairy products without appreciable performance differences (Klug and von Rymon-Lipinsky 2011).

3.3.2 Flavour Enhancers

Flavourings may be necessary in several products because of the progressive and irreversible loss of aroma intensity after technological treatments. In particular, fruit preparations suffer the remarkable reduction of flavour after heat treatment (Tamime and Robinson 1999).

Basically, flavour enhancers may be subdivided in two different categories depending on the natural or artificial origin (Tamime and Robinson 1999). Moreover, natural substances with flavouring effects might be used to replicate different aromas in relation to their botanical origin: this strategy is substantially coincident with the approach of the modern perfume industry.

Different flavours may be added to set-type or stirred-type, frozen, drinking and dried yoghurts: it should be honestly affirmed that the total list of aroma enhancers may contain thousands of names. However, regulatory restrictions and the recall to 'good manufacturing practices' may force food technologists to choose one specified and artificial compound instead of other natural molecules (Tamime and Robinson 1999).

As a result, the aroma of several flavoured yoghurts may be mainly dependent on the presence and the concentrations of following analytes (Tamime and Robinson 1999):

- 3-methylbutyl acetate, isoamyl acetate (aroma: banana fruit)
- Methyl anthranilate, 1-*p*-methene-8-thiol (aroma: grape fruits)
- Citral (aroma: lemon, orange)
- γ-decalactone (aroma: peach)
- Ethyl vanillin (aroma: vanilla).

On the other side, the aromatic profile of yoghurt products may be notably enhanced if:

- 3-methylbutyl acetate is associated with eugenol, pentyl acetate, pentyl propionate, etc. (aroma: banana fruit)
- 1-*p*-methene-8-thiol is associated with limonene, decanal, ethyl acetate, etc. (aroma: grape fruits)
- γ-decalactone is 'reinforced' with the presence of γ-octalactone, linalool, etc. (aroma: peach).

Finally, every possible aroma can be enhanced with the use of peculiar synthetic compounds: for instance, γ-undecalactone may be a good choice for peach aromas (Tamime and Robinson 1999).

In accordance with the CA, three synthetic substances can be used depending on regulatory restrictions (Codex 1995): aspartame, potassium acesulfame and neotame. It should be also considered that these compounds are mainly used as sweeteners; however, their double function may be very interesting when selected yoghurt types have to be prepared and the original—or desired—aroma has to be 'reconstituted'.

For this category of food additives, the situation of neotame is synthetically described.

This substance, also defined N-[N-(3,3-dimethylbutyl)-L-α-aspartyl]-L-phenylalanine 1-methyl ester, is obtained by aspartame and 3,3-dimethylbutyraldehyde after purification, drying and milling (Aguilar et al. 2007).

From the chemical viewpoint, the molecular formula is $C_{20}H_{30}N_2O_5$ with MW 378.47 g mol^{-1} and CAS registry number 165450-17-9 (Fig. 3.2).

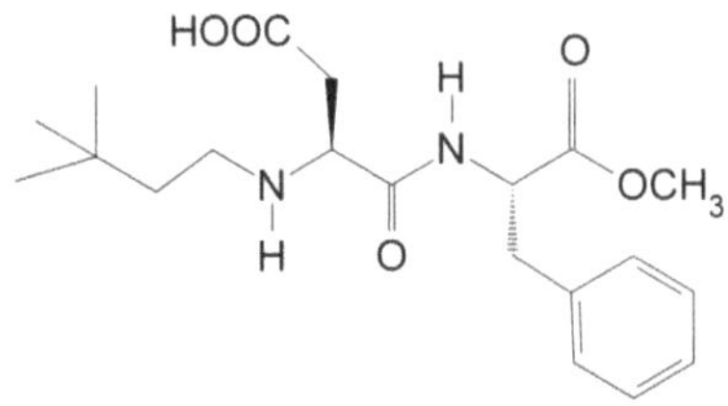

Fig. 3.2 Chemical structure of neotame, a sweetener and flavouring agent. BKchem version 0.13.0, 2009 (http://bkchem.zirael.org/index.html) has been used for drawing this structure

It can be described as a crystalline powder with possible amorphous polymorphic forms (Offerdahl et al. 2005). On the other hand, commercial forms can be also co-crystallized neotame/sugar compounds, acid or basic salts, encapsulated products, metal complexes, etc. (O'Donnell 2008). It has to be highlighted that neotame can exhibit interesting flavour properties, especially when the following aromas have to be reconstituted and/or enhanced: mint, fruit and vanilla. In these situations, the presence of neotame can justify the reduction of flavour concentrations. In addition, neotame may hide off tastes when used in conjunction with vitamins, soy and/or minerals (O'Donnell 2008).

With reference to physical features, neotame is reported to be more soluble than aspartame in water and in organic solvents such as ethanol (O'Donnell 2008). In addition, related salts should be more soluble than the normal form.

Neotame is also storable for five years at least without appreciable consequences, although high storage temperatures may cause decomposition after several months. When dissolved in water, it can be hydrolysed with the production of de-esterified neotame (O'Donnell 2008).

Generally, this additive is used in combination with other synthetic sweeteners such as saccharin and sucralose; on the other side, neotame and sweetening 'competitors' appear to show analogous performances with concern to sweetness.

3.3.3 Food Colours

The category of food additives with colouring functions has been always 'thorny' because of health suspects about the use of certain substances. Basically, the aim of food technologists should be the reconstitution of original organoleptic properties of foods with the addition of selected compounds. This approach is correct when speaking of sweeteners and flavour enhancers (Sects. 3.3.1 and 3.3.2). However, the use of food colourants in the formulation of yoghurts is mainly correlated with the necessity of attractive products (Tamime and Robinson 1999).

A potentially long list of food colourants may be shown here with an initial premise: natural and artificial colours, including derived substances, may be added with different objectives and results. In addition, coloured yoghurts should display a visual appearance in function of the claimed message: on these bases, the addition of 'banana fruit' or 'coffee' ingredients to a commercially available yoghurt 'base' might suggest the possible and non-compulsory addition of yellowlike and brownlike colours, respectively. Consequently, red cochineal may be used for strawberry-flavoured yoghurt, while tartrazine may be added to lemon-flavoured products (Calvo et al. 2001). Naturally, added colours should exhibit good miscibility and acceptable 'solidity' (resistance) against adverse conditions: thermal abuses, heat treatments, possible phase separations, excessive amount of lipids (organic phase), 'bleeding' (migration) in multilayered yoghurts, etc. (Daravingas et al. 2001).

With the exception of new natural colourants such as anthocyanin extracts by selected sources or 'old' xanthophylls like lutein dye (Carvalho et al. 2013; Domingos et al. 2014), most known food colours for yoghurts may be shown in the below-mentioned list:

- Allura red AC
- Brilliant blue FCF
- Canthaxanthin
- Caramel III
- Ammonia caramel, also named 'Caramel IV'
- Sulphite ammonia caramel
- Carmines
- Beta-carotenes (from vegetable sources)
- Synthetic beta-carotenes
- Chlorophylls (copper complexes)
- Chlorophyllin (copper complexes, potassium and sodium salts)
- Fast green FCF
- Grape skin extract
- Indigotine (indigo carmine)
- Iron oxide (black, yellow and red types)
- Ponceau 4R (also named cochineal red A)
- Riboflavin (synthetic origin)
- Riboflavin 5′-phosphate sodium
- Riboflavin by *Bacillus subtilis*
- Sunset yellow FCF.

In relation to the basic aim of this chapter, two different colourants with similar names are discussed. In detail, main features of the natural 'cochineal' and the synthetic cochineal red A are presented here.

The name 'cochineal' is historically related to a natural pigment (orange to red tints) secreted by the female exemplar of *Opuntia coccinellifera* (Wüthrich et al. 1997). The main pigment is carminic acid, molecular formula $C_{22}H_{20}O_{13}$, MW 492.38 g mol^{-1}, as shown in Fig. 3.3. It is mainly based on the central anthraquinonic structure with an additional glucose ring. Basically, it is dispersible in water; red colours appear more intense if pH increases. Unfortunately, this pigment is expensive enough at present (Downham and Collins 2000).

This molecule may be confused with 'carmine', also named 'crimson lake', 'cochineal' and 'natural red 4'. Actually, this pigment is the aluminium salt of carminic acid (Downham and Collins 2000): it shows pink to red tints. Carmine is

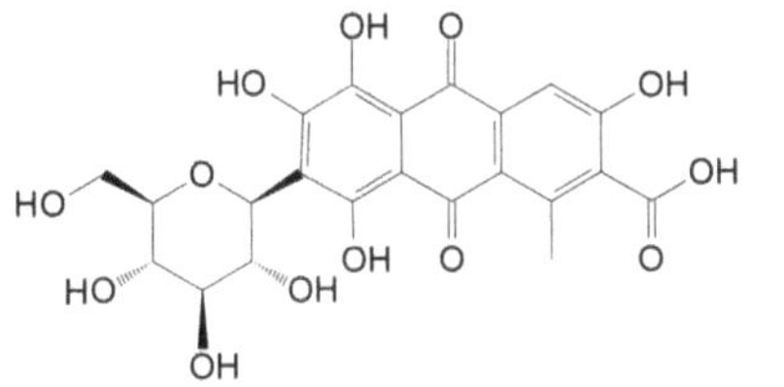

Fig. 3.3 Chemical structure of carminic acid, a pigment found in the natural cochineal secreted by the female exemplar of *Opuntia coccinellifera* (Wüthrich et al. 1997). BKchem version 0.13.0, 2009 (http://bkchem.zirael.org/index.html) has been used for drawing this structure

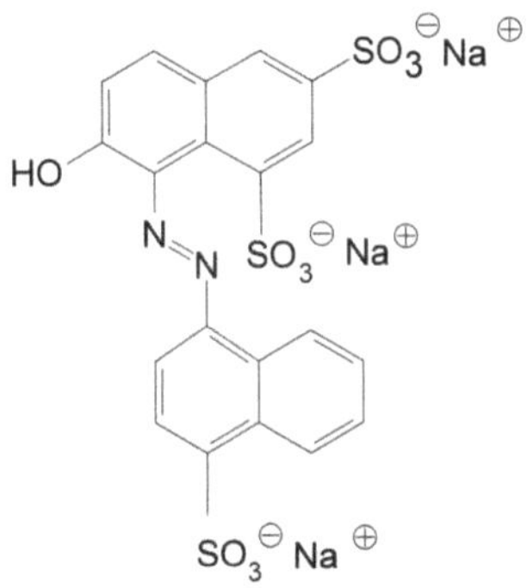

Fig. 3.4 Chemical structure of synthetic cochineal, also named ponceau 4R or cochineal red A. BKchem version 0.13.0, 2009 (http://bkchem.zirael.org/index.html) has been used for drawing this structure

used in the confectionery sector and for the production of 'non-vegetarian' foods (carminic acid is also used in this sector).

At present, carminic acid and 'cochineal' are used in the European Union and in the USA in spite of recent studies with reference to possible allergic reactions and anaphylactic shocks. The matter is currently under investigation by the European Food Safety Authority (EFSA) and the Food and Drug Administration.

The synthetic 'cochineal'—also named 'ponceau 4R', 'new coccine', 'cochineal red A', molecular formula $C_{20}H_{11}N_2Na_3O_{10}S_3$, MW 604.47 g mol^{-1}, CAS number 2611-82-7, Fig. 3.4—is a synthetic food colour. The chemical formula is completely different from carminic acid: cochineal red can be defined as trisodium 2-hydroxy-1-(4-Sulphonato-1-naphthylazo)-naphthalene-6,8-disulphonate (Aguilar et al. 2009).

Because of its strawberry red tint, it can be added to flavoured yoghurts as the only colourant or in association with other substances such as tartrazine (Lisak et al. 2012; Tamime and Robinson 1999). It can be added before or after fermentation (Mendi et al. 2004): this feature is not specific for ponceau 4R; other synthetic and natural colourants are reported to be added in the same way.

On the other hand, recent studies have highlighted the role of synthetic cochineal in association with other colouring agents in relation to the occurrence of hyperactivity in children. At present, the question is debated: the EFSA Scientific Panel on Dietetic Products, Nutrition and Allergies has concluded that the consumption of different colours—including ponceau 4R—should not cause severe adverse reactions in human subjects at the current levels of use, either individually or in combination (Agostoni et al. 2010).

Carminic acid and synthetic cochineal have been considered in relation to the whole spectrum of synthetic and natural colourants because of controversial opinions on their use. In addition, it should be noted that the use of synthetic cochineal is economically convenient if compared with natural compounds. Finally, it has been reported that ponceau 4R has good light solidity, heat and acid stability: these features are important when the use in thermally processed dairy products is proposed. On the other side, some fading may appear when synthetic cochineal is used with ascorbic acid and sulphur dioxide (Downham and Collins 2000).

3.3.4 Thickeners

This category of food additives should be considered in the general ambit of emulsifiers, sequestrants and stabilizers (Table 3.2). In fact, a little portion of emulsifiers/stabilizers may have an interesting influence on the viscosity of foods and beverages. Because of the importance of rheology, the discrimination has been operated in this section between the whole category of emulsifiers/stabilizers and 'thickening agents' (Tamime and Robinson 1999).

By a general viewpoint, normal emulsifiers are definable as amphiphilic molecules because of the concomitant presence of hydrophobic and hydrophilic groups. Consequently, these molecules can be used to promote and enhance water/oil emulsions by means of the reduction of lipidic masses in small emulsified droplets. As a result, the superficial tension is notably reduced. Should the emulsified state be maintained for extended time periods, emulsifiers would be named also stabilizers (Table 3.2). Finally, the texture of modified foods may be enhanced by means of the use of thickening agents. Consequently, the nature and physicochemical properties of these food additives can subdivide the class of 'emulsifiers' in a tripartite group depending on the final and declared use.

Table 3.2 shows a list of emulsifiers and stabilizers: sometimes, these substances may have other properties. For instance, vegetable pectins or xanthan gum may be also defined 'gelling agents' because they are able to promote the gelification of emulsified foods (Cerutti 1999). On the other hand, some of these substances may have also thickening effects: for example, the following molecules are thickeners (Tamime and Robinson 1999):

- Vegetable exudates (Arabic and tragacanth varieties)
- Vegetable seed flour (carob variety)
- Extracts from seaweeds: alginates, furcellaran
- Different cereal starches
- Cellulose derivatives

- Xanthan
- And other compounds.

According to the CA, the following list of thickeners may be used for yoghurt products and yoghurt-related foods:

- Alginates (alginic acid, ammonium alginate, calcium alginate, etc.)
- Ammonium salts of phosphatidic acid
- Diacetyltartaric and fatty acid esters of glycerol
- Calcium polyphosphate
- Ammonium polyphosphate
- Bone phosphate
- Polyoxyethylene (20) sorbitan monolaurate
- Polyoxyethylene (20) sorbitan monooleate
- Polyoxyethylene (20) sorbitan monopalmitate
- Polyoxyethylene (20) sorbitan monostearate
- Polyoxyethylene (20) sorbitan tristearate
- Propylene glycol esters of fatty acids
- Sucroglycerides.

In relation to the use of thickeners for improving rheological properties and sensorial features of yoghurts, alginates are discussed because of peculiar properties and some interesting feature in relation to the production of probiotic yoghurts.

By the chemical viewpoint, alginates are a family of unbranched binary heteropolymers containing 1,4-linked β-D-mannuronic (M) and 1,4-linked α-L-guluronic acid (G) residues with different proportions and sequences (Draget et al. 2005; Smidsrød 1974). The chain can contain two homopolymeric MM and GG blocks with the concomitant presence of mixed MG blocks (Draget et al. 2005). Anyway, the dimension of blocks appears higher for GG fragments if compared with MM blocks; in addition, MM fragments seem to be dimensionally higher than heteropolymeric MG blocks (Smidsrød 1974).

The main property of alginates is correlated with the selective binding of calcium ions in solution for GG blocks; moreover, this chemical phenomenon is maintained during time (Smidsrød 1974). On the other hand, MM and MG blocks do not appear to show good selectivity for calcium ions, auto-cooperative binding mechanisms and recognizable hysteresis (Smidsrød 1974). As a result, thickening properties appear to be mainly caused by the abundance of GG fragments in calcium alginate gels (Smidsrød 1974). However, it has been also reported that alginates have not regular statistical distributions of different blocks (Draget et al. 2005). Actually, one main difference may be observed on a molecular scale between two different types of available alginates because of the origin: bacterial products seem to show O-acetyl groups in C2 and/or in C3 position along the chain if compared with algal alginates (Draget et al. 2005).

With concern to physical properties, alginates appear to be more interesting than other polysaccharides because of the selective binding of multivalent ions; in addition, the reported sol/gel transition of alginates appears to be independent from thermal modifications (Draget et al. 2005).

On the other hand, alginates may be dissolved with some difficulty depending on the pH of solvents and the resulting influence on electrostatic charges. Moreover, calcium and other multivalent ions (e.g. magnesium) should be abundant in comparison with other non-gelling ions (Haug and Smidsrød 1965). Otherwise, the thickening or gelling effects could be insufficient (Draget et al. 2005). Actually, these phenomena appear similar on the macroscopic level.

Another reflection should be made with reference to pH. In fact, a controlled and slow decrease of pH can favour the formation of alginic acid gels, while the sudden diminution of the proton concentration below known pKa values may produce the precipitation of alginate molecules without binging and the consequent thickening action. For this reason, the use of propylene glycol alginate may be recommended as a food stabilizer (Draget et al. 2005; Xiaoying et al. 2009).

In addition, the contemporary presence of multivalent ions in notable quantity produces often rapid and irreversible binding reactions with undesired heterogeneous and irregular gels (Draget et al. 1990).

With reference to the use of alginates for yoghurt productions, it should be also remembered that these polymers can be easily depolymerized by oxidative–reductive reactions, depending on the pH and temperature. The depolymerization should be taken into account when heat treatments are planned (Draget et al. 2005).

Finally, the use of alginates has been proposed in an innovative way because of the necessity of increasing the survival and viability of probiotic bacteria in yoghurt during storage. Generally, the use of calcium-induced alginate–starch-encapsulated probiotic bacteria has been proposed and studied. Obtained results seem to demonstrate that similar strategies do not affect sensory properties of produced yoghurts (Grosso and Fávaro-Trindade 2004; Kailasapathy 2006; Krasaekoopt et al. 2006; Sultana et al. 2000).

3.4 The Influence of Food Additives on the Design of Yoghurt

As mentioned above, the use of food additives for the production of modern yoghurts is necessary when certain properties or positive features have to be obtained (Sect. 3.3). Because of the influence of marketing strategists and consumers on the success (or the commercial failure) of every consumer good, several properties are substantially 'implicit' when speaking of modern—plain, flavoured, coloured and drinking—yoghurts and dairy-based desserts. As a result, one or more of discussed chemicals or classes of food additives are needed with the aim of assuring ab initio the following properties:

- Increased viscosity, when needed; drinking products should appear 'diluted' in comparison with plain yoghurts
- Ameliorated 'sweet effect'; the sweeter the product, the higher the acceptability for normal consumers

- Augmented aroma of the final product
- Chromatic performance of the fluid composition and absence of 'bleeding' effects (Sect. 3.3.3). The intensity of obtained colours should not be modified throughout the whole shelf life.

By a general viewpoint, these priority features are directly correlated (and possibly measurable) with organoleptic testing methods. Consequently, this chapter has discussed four categories of food additives: colourants, sweeteners, flavouring agents and thickeners, while remaining classes have been only mentioned. The choice has been determined by the following considerations:

- When deciding the best approach for the production of commercially attractive products, basic sensorial features are critical and absolutely 'urgent'
- Colour, flavour and taste are immediately measurable by normal consumers (Parisi 2012)
- On these bases, the choice of food additives has to take into account the necessity of reconstituting pre-existing or supposed aroma, colour and taste of original yoghurts and claimed ingredients. Naturally, the creation of a peculiar aroma (with correlated chromatic 'codes' and corresponding tastes) may be tried if the formulation does not include 'natural' food ingredients. Anyway, the expectation of the normal consumer has to be confirmed
- In addition, every product has a recognizable aspect and a peculiar texture. Once more, these features have to be 'replicated' for every new version or subversion of the 'original' prototype or traditional food (Parisi 2012). For this reason, the use of peculiar emulsifiers, stabilizers and/or thickeners is generally requested in the modern industry. Otherwise, the risk of irregular products may depend on the variability of raw materials, packaging materials and processing parameters. Normal examples may concern the so-called bleeding effect in multilayered products (Sect. 3.3.3), the presence of irregular gels or phase separations with possible consumer complaints, abnormal fermentations, etc.

In conclusion, the problem of food stability should be studied and solved on the basis of the preliminary design—the choice of ingredients, flavour enhancers, colours, sweeteners, emulsifiers, etc.—by the viewpoint of the food chemist. In fact, obtained products should remain stable throughout the whole shelf life: this feature implies the absence or the limitation of phase separations in the product. In addition, above-mentioned features and other technological properties (e.g. sprayability, fluidity before use) have to be obligatorily constant when yoghurts are used as ingredients for other food products.

The stability of prepared yoghurts may require the additional use of substances such as antioxidants (ascorbic acid, carotenoids, etc.), anticaking agents and preservatives (benzoic acid, potassium sorbate, etc.). These compounds are used with the aim of increasing the microbiological and commercial shelf life of produced yoghurts. In other words, packaged foods have to maintain their own physicochemical and microbiological features until the end of the declared expiration date. Because of the priority importance of sensorial features, all described additives in Tables 3.1 and 3.2 have not been mentioned here in detail. The author would discuss these chemicals and related functions in a future issue of this book series in Chemistry of Foods.

References

Abu-Jdayil B, Shaker RR, Jumah RY (2000) Rheological behavior of concentrated yogurt (Labneh). Int J Food Properties 3(2):207–216. doi:10.1080/10942910009524628

Agostoni C, Bresson JL, Fairweather-Tait S, Flynn A, Golly I, Korhonen H, Lagiou P, Løvik M, Marchelli R, Martin A, Moseley B, Neuhäuser-Berthold M, Przyrembel H, Salminen S, Sanz Y, Strain SJJ, Strobel S, Tetens I, Tomé D, van Loveren H, Verhagen H (2010) Scientific Opinion on the appropriateness of the food azo-colours Tartrazine (E 102), Sunset Yellow FCF (E 110), Carmoisine (E 122), Amaranth (E 123), Ponceau 4R (E 124), Allura Red AC (E 129), Brilliant Black BN (E 151), Brown FK (E 154), Brown HT (E 155) and Litholrubine BK (E 180) for inclusion in the list of food ingredients set up in Annex IIIa of Directive 2000/13/EC1. EFSA J 8(10):1778. doi:10.2903/j.efsa.2009.1328

Aguilar F, Autrup H, Barlow S, Castle L, Crebelli R, Dekant W, Engel KH, Gontard N, Gott D, Grilli S, Gürtler R, Larsen JC, Leclercq C, Leblanc JC, Malcata FX, Mennes W, Milana MR, Pratt I, Rietjens I, Tobback P, Toldrá F (2007) Neotame as a sweetener and flavour enhancer. Scientific opinion of the panel on food additives, flavourings, processing aids and materials in contact with food. EFSA J 581:1–43

Aguilar F, Charrondiere UR, Dusemund B, Galtier P, Gilbert J, Gott DM, Grilli S, Guertler R, Koenig J, Lambré C, Larsen JC, Leblanc JC, Mortensen A, Parent-Massin D, Pratt I, Rietjens IMCM, Stankovic I, Tobback P, Verguieva T, Woutersen RA (2009) Scientific opinion on the re-evaluation of Ponceau 4R (E 124) as a food additive. EFSA J 7(11):1328. doi:10.2903/j.efsa.2009.1328

Aichinger PA, Michel M, Servais C, Dillmann ML, Rouvet M, D'Amico N et al (2003) Fermentation of a skim milk concentrate with Streptococcus thermophilus and chymosin: structure, viscoelasticity and syneresis of gels. Coll Surf B: Biointerfaces 31(1):243–255. doi:10.1016/S0927-7765(03)00144-9

Akpan UG, Mohammed AD, Aminu I (2007) Effect of preservative on the Shelf life of Yoghurt produced from soya beans milk. Leonardo Electron J Pract Technol 6(11):131–142

Alichanidis E, Polychroniadou A (1996) Special features of dairy products from ewe and goat milk from the physicochemical and organoleptic point of view. In: Proceedings, IDF seminar, production and utilization of EWE and goat milk, Crete, Greece, 19–21 Oct 1995, Bulletin no. 202. International Dairy Federation, Brussels, pp 21–43

Altman L, Landis W (1995) Acceptability of common muffin and cake recipes using non-fat plain yogurt as a fat replacement. J Am Diet Ass 1(4):45–54. doi:10.1016/S0002-8223(95)00490-4

American Dietetic Association (2004) Position of the American dietetic association: use of nutritive and nonnutritive sweeteners. J Am Diet Assoc 104(2):255–275

Anifantakis EM (1986) Comparison of the physico-chemical properties of ewes' and cow' milk. In: Proceedings, IDF seminar, production and utilization of Ewe's and Goat's milk, Athens, Greece, 23–25 September 1985, Bulletin No. 202. International Dairy Federation, Brussels, pp 42–53

Aryana KJ (2003) Folic acid fortified fat-free plain set yoghurt. Int J Dairy Technol 56(4):219–222. doi:10.1046/j.1471-0307.2003.00105.x

Basak S, Ramaswamy HS (1994) Simultaneous evaluation of shear rate and time dependency of stirred yogurt rheology as influenced by added pectin and strawberry concentrate. J Food Eng 21(3):385–393. doi:10.1016/0260-8774(94)90081-7

Benezech T, Maingonnat JF (1994) Characterization of the rheological properties of yoghurt—a review. J Food Eng 21(4):447–472. doi:10.1016/0260-8774(94)90066-3

Borneo R, Kocer D, Ghai G, Tepper BJ, Karwe MV (2007) Stability and consumer acceptance of long-chain omega-3 fatty acids (Eicosapentaenoic Acid, 20:5, n-3 and Docosahexaenoic Acid, 22:6, n-3) in cream-filled sandwich cookies. J Food Sci 72(1):S049–S054. doi:10.1111/j.1750-3841.2006.00240.x

Calvo C, Salvador A, Fiszman SM (2001) Influence of colour intensity on the perception of colour and sweetness in various fruit-flavoured yoghurts. Eur Food Res Technol 213(2):99–103. doi:10.1007/s002170100359

Carvalho Alves AP, Corrêa AD, Pinheiro ACM, Oliveira FC (2013) Flour and anthocyanin extracts of jaboticaba skins used as a natural dye in yogurt. Int J Food Sci Technol 48(10):2007–2013. doi:10.1111/ijfs.12110

Cerutti G (1999) Residui, additivi e contaminanti degli alimenti. Tecniche nuove, Milan

Chen MJ, Jovanovic A, Taylor R (2010) Utilizing the second-meal effect in type 2 diabetes: practical use of a soya-yogurt snack. Diabet Care 33(12):2552–2554. doi:10.2337/dc10-0552

Codex (1995) Codex general standard for food additives. STAN 192-1995. Joint FAO/WHO Food Standards Programme, FAO, Rome

Codex (2012) CODEX standard for fermented milks. Codex Stan 243-2003. Food and Agriculture Organization, United Nations, Rome, pp 1–5

Collet LSFCA, Tadini CC (2004) Sodium Caseinate addition effect on the thixotropy of stirred yogurt. In: Proceedings of international congress on engineering and food iCEF vol 9, Montpellier, pp 317–322

Daravingas GV, Heitke TC, Funk DF (2001) Colored multi-layered yogurt and methods of preparation. US Patent 6,235,320, 22 May 2001

Domingos LD, Xavier AAO, Mercadante AZ, Petenate AJ, Jorge RA, Viotto WH (2014) Oxidative stability of yogurt with added lutein dye. J Dairy Sci 97(2):616–623. doi:10.3168/jds.2013-6971

Downham A, Collins P (2000) Colouring our foods in the last and next millennium. Int J Food Sci Technol 35(1):5–22. doi:10.1046/j.1365-2621.2000.00373.x

Draget KI, Østgaard K, Smidsrød O (1990) Homogeneous alginate gels: a technical approach. Carb Poly 14(2):159–178. doi:10.1016/0144-8617(90)90028-Q

Draget KI, Smidsrød O, Skjåk-Bræk G (2005) Alginates from algae. In: Steinbüchel A, Rhee SK (eds) Polysaccharides and polyamides in the food industry. Properties, production and patents. Wiley-VCH Verlag GmbH & Co, Weinheim. doi:10.1002/3527600035.bpol6008

Glinsmann WH, Irausquin H, Park YK (1986) Evaluation of health aspects of sugars contained in carbohydrate sweeteners. Report of Sugars Task Force, 1986. J Nutr 116(11):S1–216

Grosso CRF, Fávaro-Trindade CS (2004) Stability of free and immobilized Lactobacillus acidophilus and Bifidobacterium lactis in acidified milk and of immobilized B. lactis in yoghurt. Braz J Microbiol 35(1–2):151–156. doi:10.1590/S1517-83822004000100025

Guggisberg D, Cuthbert-Steven J, Piccinali P, Bütikofer U, Eberhard P (2009) Rheological, microstructural and sensory characterization of low-fat and whole milk set yoghurt as influenced by inulin addition. Int Dairy J 19(2):107–115. doi:10.1016/j.idairyj.2008.07.009

Harwalkar VR, Kalab M (1986) Relationship between microstructure and susceptibility to syneresis in yogurt made from reconstituted nonfat dry milk. Food Microstruct 5:287–294

Haug A, Smidsrød O (1965) The effect of divalent metals on the properties of alginate solutions. Acta Chem Scand 19(2):341–345

Jaziri I, Ben Slama M, Mhadhbi H, Urdaci MC, Hamdi M (2009) Effect of green and black teas (Camellia sinensis L.) on the characteristic microflora of yogurt during fermentation and refrigerated storage. Food Chem 112(3):614–620. doi:10.1016/j.foodchem.2008.06.017

Kailasapathy K (2006) Survival of free and encapsulated probiotic bacteria and their effect on the sensory properties of yoghurt. LWT-Food Sci Technol 39(10):1221–1227. doi:10.1016/j.lwt.2005.07.013

Kälviäinen N, Roininen K, Tuorila H (2003) The relative importance of texture, taste and aroma on a yogurt-type snack food preference in the young and the elderly. Food Qual Prefer 14(3):177–186. doi:10.1016/S0950-3293(02)00049-6

Kaminarides S, Stamou P, Massouras T (2007) Comparison of the characteristics of set type yoghurt made from ovine milk of different fat content. Int J Food Sci Technol 42(9):1019–1028. doi:10.1111/j.1365-2621.2006.01320.x
Keating KR, White CH (1990) Effect of alternative sweeteners in plain and fruit-flavored yogurts. J Dairy Sci 73(1):54–62. doi:10.3168/jds.S0022-0302(90)78645-6
Kiani H, Mousavi SMA, Emam-Djomeh Z (2008) Rheological properties of Iranian yoghurt drink, Doogh. Int J Dairy Sci 3(2):71–78
Klug C, von Rymon Lipinsky GW (2011) Acesulfame potassium. In: O'Brien Nabors L (ed) Alternative sweeteners, vol 48. CRC Press, Boca Raton
Koç B, Yilmazer MS, Balkır P, Ertekin FK (2010) Moisture sorption isotherms and storage stability of spray-dried yogurt powder. Drying Technol 28(6):816–822. doi:10.1080/07373937.2010.485083
Krasaekoopt W, Bhandari B, Deeth HC (2006) Survival of probiotics encapsulated in chitosan-coated alginate beads in yoghurt from UHT-and conventionally treated milk during storage. LWT-Food Sci Technol 39(2):177–183. doi:10.1016/j.lwt.2004.12.006
Kurmann JA (1986) Yogurt made from ewe and goats milk. In: Proceedings, production and utilization of Ewe's and Goat's milk, Athens, Sep. 23–25 September 1985. International Dairy Federation Bulletin 202, pp 153–166
Lisak K, Lenc M, Jeličić I, Božanić R (2012) Sensory evaluation of the strawberry flavored yoghurt with stevia and sucrose addition. Croatian J Food Technol Biotechnol Nutr 7:39–43. http://hrcak.srce.hr/file/123163. Accessed 25 Mar 2014
Maga JA, Tu AT (1995) Food additive toxicology. Marcel Dekker Inc, New York
Magenis RB, Prudêncio ES, Amboni RD, Cerqueira Júnior NG, Oliveira RV, Soldi V, Benedet HD (2006) Compositional and physical properties of yogurts manufactured from milk and whey cheese concentrated by ultrafiltration. Int J Food Sci Technol 41(5):560–568. doi:10.1111/j.1365-2621.2005.01100.x
Mendi SD, Peters B, Kamga PT (2004) The effects of pH and heat treatment processing on the stability of natural food colours used in dairy products. J Food Technol Afr 5(2):59–61
Modler HW, Kalab M (1983) Microstructure of yogurt stabilized with milk proteins. J Dairy Sci 66(3):430–437. doi:10.3168/jds.S0022-0302(83)81810-4
Mohameed HA, Abu-Jdayil B, Al-Shawabkeh A (2004) Effect of solids concentration on the rheology of labneh (concentrated yogurt) produced from sheep milk. J Food Eng 61(3):347–352. doi:10.1016/S0260-8774(03)00139-0
Najgebauer-Lejko D (2014) Effect of green tea supplementation on the microbiological, antioxidant, and sensory properties of probiotic milks. Dairy Sci Technol 1–13. doi:10.1111/ijfs.12411
Nahar A, Al-Amin M, Alam SMK, Wadud A, Islam MN (2007) A comparative study on the quality of Dahi (Yoghurt) prepared from cow, goat and buffalo milk. Int J Dairy Sci 2(3):260–267
O'Brien Nabors L (2011) Alternative sweeteners: an overview. In: O'Brien Nabors L (ed) Alternative sweeteners, vol. 48. CRC Press, Boca Raton
O'Donnell K (2008) Aspartame and Neotamein. In: Mitchell H (ed) Sweeteners and sugar alternatives in food technology. Blackwell Publishing, Oxford, pp 86–102
Offerdahl TJ, Salsbury JS, Dong Z, Grant DJ, Schroeder SA, Prakash I, Gorman EM, Barich DH, Munson EJ (2005) Quantitation of crystalline and amorphous forms of anhydrous neotame using 13C CPMAS NMR spectroscopy. J Pharm Sci 94(12):2591–2605. doi:10.1002/jps.20469
Parisi S (2012) Food packaging and food alterations. Smithers Rapra, Shawbury
Park HW, Chung HJ, Choi EJ, Lee JJ (2002) The development of high fiber food for constipation. Korean J Community Nutr 7(5):715–723
Penna ALB, Sivieri K, Oliveira MN (2001) Relation between quality and rheological properties of lactic beverages. J Food Eng 49(1):7–13. doi:10.1016/S0260-8774(00)00179-5
Peterson MA (1979) Yogurt flavored confectioneries. US Patent 4,150,163, 17 Apr 1979
Prakash I, DuBois GE, Clos JF, Wilkens KL, Fosdick LE (2008) Development of rebiana, a natural, non-caloric sweetener. Food Chem Toxic 46(7):S75–S82. doi:10.1016/j.fct.2008.05.004
Rowe RC, Sheskey PJ, Weller PJ (eds) (2003) Handbook of pharmaceutical excipients, vol 4. Pharmaceutical Press, London

Schiffman SS, Sattely-Miller EA, Graham BG, Zervakis J, Butchko HH, Stargel WW (2003) Effect of repeated presentation on sweetness intensity of binary and ternary mixtures of sweeteners. Chem Senses 28(3):219–229. doi:10.1093/chemse/28.3.219

Schlotke F, Becker W, Ireland J, Møller A, Ovaskainen ML, Monspart J, Unwin I (2000) EUROFOODS recommendations for food composition database management and data interchange. J Food Compos Anal 13(4):709–744. doi:http://dx.doi.org/10.1006/jfca.2000.0891

Sendra E, Kuri V, Fernández-López J, Sayas-Barberá E, Navarro C, Pérez-Alvarez JA (2010) Viscoelastic properties of orange fiber enriched yogurt as a function of fiber dose, size and thermal treatment. LWT-Food Sci Technol 43(4):708–714. doi:10.1016/j.lwt.2009.12.005

Serrano FO, López IS, Revilla GN (1991) High performance liquid chromatography determination of chemical preservatives in yogurt. J Liq Chrom 14(4):709–717. doi:10.1080/01483919108049282

Smidsrød O (1974) Molecular basis for some physical properties of alginates in the gel state. Faraday Discuss Chem Soc 57:263–274. doi:10.1039/DC9745700263

Smith J (ed) (1991) Food additive user's handbook. Blackie, London

Sultana K, Godward G, Reynolds N, Arumugaswamy R, Peiris P, Kailasapathy K (2000) Encapsulation of probiotic bacteria with alginate—starch and evaluation of survival in simulated gastrointestinal conditions and in yoghurt. Int J Food Microbiol 62(1):47–55. doi:10.1016/S0168-1605(00)00380-9

Tamime AY, Robinson RK (1999) Biochemistry of fermentation. In: Tamime AY, Robinson RK (eds) Yoghurt science and technology, 2nd edn. CRC Press, Cambridge

Truswell AS, Bateson D, Madafiglio D, Pennington JAT, Rand WM, Klensin JC (1991) INFOODS guidelines for describing foods to facilitate international exchange of food composition data. J Food Compos Anal 4(1):18–38. doi:10.1016/0889-1575(91)90045-8

Wüthrich B, Kägi MK, Stücker W (1997) Anaphylactic reactions to ingested carmine (E120). Allergy 52(11):1133–1137. doi:10.1111/j.1398-9995.1997.tb00189.x

Xiaoying W, Quanyang L, Hongling Z, Zhengtao Z (2009) Impact of propylene Glgcol alginate on yogurt. Food Ferment Ind 2:066

Younus S, Masud T, Aziz T (2002) Quality evaluation of market yoghurt/dahi. Pak J Nutr 1(5):226–230

Zanjani MAK, Tarzi BG, Sharifan A, Mohammadi N, Bakhoda H, Madanipour MM (2012) Microencapsulation of Lactobacillus casei with calcium alginate-resistant starch and evaluation of survival and sensory properties in cream-filled cake. Afr J Microbiol Res 6(26):5511–5517. doi:10.5897/AJMR12.972

Zhengtao Z, Quanyang L, Xiuju W, Xin A, Chuansen Z (2009) The influence of zymogens and xanthan on the rheology properties of yogurt. J Dairy Sci Technol 2:006

MIX
Papier aus verantwortungsvollen Quellen
Paper from responsible sources
FSC® C105338

If you have any concerns about our products,
you can contact us on
ProductSafety@springernature.com

In case Publisher is established outside the EU,
the EU authorized representative is:
Springer Nature Customer Service Center GmbH
Europaplatz 3, 69115 Heidelberg, Germany

Printed by Libri Plureos GmbH
in Hamburg, Germany